T0177550

CONTESTED KNOWLEDGE

CONTESTED KNOWLEDGE
Science, Media, and Democracy in Kerala

SHIJU SAM VARUGHESE

OXFORD
UNIVERSITY PRESS

OXFORD
UNIVERSITY PRESS

Oxford University Press is a department of the University of Oxford.
It furthers the University's objective of excellence in research, scholarship,
and education by publishing worldwide. Oxford is a registered trademark of
Oxford University Press in the UK and in certain other countries.

Published in India by
Oxford University Press
YMCA Library Building, 1 Jai Singh Road, New Delhi 110 001, India

First Edition published in 2017

ISBN-13: 978-0-19-946912-3
ISBN-10: 0-19-946912-1

Typeset in ITC Giovanni Std 9.5/13
by The Graphics Solution, New Delhi 110092
Printed in India by Rakmo Press, New Delhi 110 020

Contents

Foreword

As long as a triumphant science was a core element of the imaginary of science, historical studies of science and society, amongst other problematics and research themes, centred the popularization and the transmission of so-called modern science. These histories of the expansion of the dominion of modern science were premised on a master narrative where as journey and process science was a movement towards the truth, that en route dislodged or eliminated irrational and other ways of knowing and acting on the world. As a journey towards the truth it was endowed with a moral value, without necessarily having its origins in a sacred cosmology. Within studies on the history of sciences, scientific knowledge as a set of travelling theories, practices, and institutional forms, detailed attention was paid to the reception of a widely accepted or revolutionary theory. These studies, till the 1950s, within multiple contexts of national research systems investigated the social and epistemic context of conversion, but by the 1970s began to look at the interlocking and dialectic between 'other ways of knowing' and the 'forward' march of science. In India too, the historicization of the so-called traditional sciences and modern sciences was framed by the idea of the stabilization of the identity of modern science and reception studies were more or less pinioned on an epistemology of the sciences that one might consider transcendent.

However, over the last three decades of the last century, as the stranglehold of a positivist conception of science weakened, a number of disciplinary fields and influences, ranging from the philosophy of science and the sociology of knowledge to anthropology and critiques of development studies, posed severe challenges for the master narrative of science that informed a new genre of studies on science and society, in the process reproblematizing earlier research themes along several axes such as science and the public, science and the media, science and popular literature, or science and literature. This reproblematization was also inspired by the cultural turn in science studies that deconstructed the notion of the popular and the public. The task was then to contextualize the diversity of differently structured communities and communicative practices, highlighting in turn the different asymmetries marking the producers and consumers of scientific knowledge, endowing the so-called consumers with some active agency in the domestication of the cultures of science.

But the institutional and intellectual linkages—rarely acknowledged, as pointed out by historians of science—between the practice of science and the concerns of scholars were reflected in a growing antagonism between the world of science and scholarly studies on science and society. In global terms, since the late 1980s scientific communities found the need to respond to the declining enrolment in the sciences. In academic terms one of the diagnostic responses was the ascent of the research theme of the public understanding of science, initially devoted to the task of measuring the public understanding of science. This was to inform policy bodies and state funding agencies sponsoring further projects on the popularization of science, including science writing. This early movement was quickly followed by the contextualist turn in the public understanding of science that took upon itself the task to engage with the dialectic between common knowledge and modern science, as well as with differently institutionally inscribed notions of expertise and its legitimation.

In India's small science studies circle, Shiju Sam Varughese's thesis on science and the media belonged to the early studies on the subject within the contextualist frame. In studying the process of negotiation between different kinds of expert knowledge, Varughese has himself become an expert on the subject. The present book, which has been reworked, revisited, and rewritten, reflects a theoretical maturity in thinking through the relationship between science, media, and the public. This relationship could be visualized in terms of a semiotic triangle represented by institutional science (seen as the generator of a particular kind of knowledge), the media (not just as a communicator of this knowledge to the public, but as the lens through which different agendas of a variety of actors are brokered), and very specific public(s) where science impacts and where science is reshaped by other knowledge(s), and is filtered through the lens of both the media and public. Varughese has exemplified these relationships through concrete studies where public controversies involving science have erupted over the last decades in Kerala. Evidently, he has attempted to situate these controversies not just within science studies, but within Kerala's social history, in resonance with the social theoretic idea of regional modernities.

Anthropologically speaking, we could argue that earlier studies on the processes of scientization and modernization were framed in a limited way through the prism of mimetic processes. Varughese's work on the other hand looks at these public controversies through the processes of the staging and enactment of encounters with big science and 'other knowledges', thus raising the question of the performativity of science in different institutional spheres discussed in the book. This could potentially push the studies of science and society towards the little explored possibilities of the sociology of convention and ritual. Perhaps that is not a direction that Varughese wishes to venture into. It is not my purpose to summarize the author's argument, but merely to locate it within the landscape of studies on science and society, and to highlight the significant interpretive and social theoretic moments in the contextualist public understanding of science. Varughese has indeed taken care to acquire some analytical and theoretical distance from the problematic identified and this is most certainly a very important work in the contextualist understanding of 'science and the public'.

Dhruv Raina
Jawaharlal Nehru University
New Delhi

Acknowledgements

L et me take this opportunity to express my sincere gratitude to those who have influenced in various capacities the research that culminated in this book. Initial conceptualization of the project took shape as part of my doctoral research at Jawaharlal Nehru University (JNU), New Delhi, between 2003 and 2009. I am heavily indebted to Dhruv Raina for his unconditional support and mentorship as my thesis supervisor. I am grateful for his unabated patience in grooming and disciplining me as a researcher, and for our deep friendship. He was also kind enough to write a foreword to the book. Rajeswari S. Raina and S. Irfan Habib also played important roles in shaping the work. Their intellectual and emotional support and friendship have been of great importance to me. Abhinav Raina grew up from a six-year-old into a college boy and musician in front of my eyes during the last thirteen years I spent on this project, and I thank him for his affection and music.

I am obliged to Dilip M. Menon, Helene E. Longino, J. Devika, Mary E. John, Nasir Tyabji, and Prajit K. Basu for their critical interventions at different stages of the project, which immensely helped me prepare the manuscript, although many of them might not even remember the same. Comments and suggestions from the two unknown readers of the manuscript at Oxford University Press are gratefully acknowledged. The long and often heated discussions with Biswanath Dash, George Varghese, Justin Mathew, and Mathew George helped me realize the weaknesses of my own understanding of the research problem. The close reading of an earlier draft of the manuscript by Arathi P.M., Rachel Varghese, and Sreejith Divakaran as a team was immensely helpful in revising the arguments. Thanks are also due to Ashok R. Chandran for his critical examination of my earlier research on science and media in his column, 'Lighthouse', in *Media* (vol. 2, no. 7, November 2013, pp. 38–9), a bilingual monthly journal of the Kerala Press Academy. AKG Centre Library (Thiruvananthapuram, Kerala) offered me its bountiful collection of newspapers and magazines which made the archival work easier. (Now late) Com. Andalattu (librarian), Ms Girija, and Ms Komalam (assistant librarians) were always ready to offer a helping hand. Special thanks to P.V. Aniyan for introducing me to the library. George Zachariah, Jessica Wang, and Satheese Chandra Bose helped me procure some important secondary sources. The scientists, journalists, and social activists whom I interviewed were highly cooperative, and enthusiastically responded to my queries. Without their help, this research would have been impossible.

Many of the arguments of the book were initially rehearsed in classrooms as part of the MPhil-PhD courses, 'Science Communication: Approaches and Methods', 'Introduction to Science, Technology and Society (STS) Studies', and 'Science and Technology in Modern India', at the Centre for Studies in Science, Technology and Innovation Policy of the Central University of Gujarat. I thank the participants of these courses. Interaction with my research students—Abhishek V. Lakkad, Kanika Sharma, Mehul Chavada, Pankajkumar J. Vaghela, Subair K., Vikram Singh Brijwal, and Vinod Kumar Singh—helped me think through the project, and I thank them for their enthusiasm and intellectual companionship.

The Kerala Modernity Studies Collective, of which I am a member, has a deep influence on shaping my thoughts. I thank the fellow members of the collective—Abhilash Malayil, Ameet Parameswaran, Arathi P.M., Aneish P. Rajan, Justin Mathew, Rachel Varghese, Satheese Chandra Bose, and Vinod Kottayil Kalidasan—for their camaraderie. Bandar Collective, the discussion group at JNU, re-contoured my intellectual orientation; thanks to its members.

Thanks are also due to Anshi David, Anish P. Raveendran, Anup Sam Ninan, (late) Bandranglila Jamir, Biju P. John, Deepak Kumar, Elias Jarso, E.V. Ramakrishnan, George Zachariah, Jose Peter, Jossy Kurien, Krishnakumar M.V., Mathew K., Oommen V. Varkey, Philip Mathew, Prabha Mathew, Pramod N., Pramod Shankar, Rosmin Mathew, Saji M., Sam Koshy, Santhosh George, Saroj Bangaru Laxman, Sarosh Koshy,

Saumen Chattopadhyay, Saleena P., Sheeba Mathew, Siby Easow, Sonia George, Sukumaran C.V., Sunny George, Tiby Mathew, Tina Thomas, Tresa Benjamin, Y. Das, Uma Purushothaman, and Yacob Thomas for their friendship and unconditional support at various stages of my research. Achuit Singh, Anushka Gokhale, Archi Haldania, Asima Jena, Atul Misra, Charu Dube, Madhumita Biswal, Namrata Kilpady Mishra, Navaneetha Mokkil, Rajesh Vasita, Saurabh Sharma, Urmila Bhirdikar, Vaibhav Abnave, and Vijendra Singh, friends in Gandhinagar, made my life easier and meaningful during the last six years. Thank you, dear friends.

A small part of Chapter 5 was previously published in *Spontaneous Generations: A Journal of History and Philosophy of Science* (vol. 5, no.1, 2011, pp. 36–43) under the title 'Media and Public Controversies over Science: A Case from Kerala, India'. A much earlier version of Chapter 6 appeared under the title '"It's Raining Aliens!" Coloured Rain in Kerala and the Fuzzy Frontiers of Science' in *Frontiers: Sarai Reader 2007* (edited by Monica Nerula et al., The Sarai Programme, CSDS, Delhi, pp. 348–58). I thank both the open access publications and their anonymous reviewers. A preliminary draft of Chapter 5 was read at the Sarai/CSDS Independent Fellowship Seminar, New Delhi, in December 2007. An earlier draft of Chapter 3 was presented at a seminar organized by the Dr Thomas Mar Athanasius Memorial Research and Orientation Centre, Kottayam, Kerala in June 2009. A section of Chapter 5 was presented at the twentieth conference of Gujarat Sociological Society at the Central University of Gujarat, Gandhinagar, in January 2013. Chapters 2 and 4 were read at public lectures organized by the Homi Bhabha Centre for Science Education, Mumbai, in February 2016. I am grateful to the discussants and participants of these seminars for their valuable comments and suggestions.

My sincere thanks to the editorial team at the Oxford University Press for the professional assistance extended towards the successful publication of the book.

I thank Amma, Pappa, Liju, and Julie for holding me close to their hearts and unconditionally supporting my journeys. Jomy Abraham offered strong intellectual and emotional support throughout the years I brooded over the manuscript—a big hug to her.

Abbreviations

AIDS	acquired immune deficiency syndrome
AIYF	All India Youth Federation
ANT	actor-network theory
BBC	British Broadcasting Corporation
CDS	Centre for Development Studies
CESS	Centre for Earth Science Studies
CITU	Centre of Indian Trade Unions
CM	chief minister
CPI(M)	Communist Party of India (Marxist)
CRO	contract research organization
CSMRS	Central Soil and Materials Research Station
CUSAT	Cochin University of Science and Technology
DCGI	Drug Controller General of India
DMG	Department of Mining and Geology
DMK	Dravida Munnetra Kazhakam
DNA	deoxyribonucleic acid
DST	Department of Science and Technology
DYFI	Democratic Youth Federation of India
FDA	Food and Drug Administration
FP	family planning
FPP	Family Planning Programme
FPR	Forum for Patients' Rights
G4N	tetraglycinal nor-dihydro-guairetic acid
GoI	Government of India
GSI	Geological Survey of India
HRC	Human Rights Commission

ICMR	Indian Council of Medical Research
IMA	Indian Medical Association
IMD	India Meteorological Department
IPA	Indian Pharmaceutical Association
ISRO	Indian Space Research Organisation
IUD	intrauterine device
JHU	Johns Hopkins University
JNU	Jawaharlal Nehru University
KSEB	Kerala State Electricity Board
KSGD	Kerala State Groundwater Department
KSSP	Kerala Sasthra Sahithya Parishad
M4N	tetra-O-methyl nor-dihydro-guairetic acid
Mal.	Malayalam
NDGA	nor-dihydro-guairetic acid
NGO	non-governmental organization
NGRI	National Geophysical Research Institute
NISTADS	National Institute of Science, Technology and Development Studies
PEST	public engagement with science and technology
PIL	public interest litigation
PRO	public relations officer
PSM	people's science movement
PUS	public understanding of science
RCC	Regional Cancer Centre
RIS	reservoir-induced seismicity
RNA	ribonucleic acid
S&T	science and technology
SALS	School of Applied Life Sciences
SPAP	School of Pure and Applied Physics
SSK	sociology of scientific knowledge
STS	science, technology, and society
TBGRI	Tropical Botanical Garden and Research Centre
TV	television
UK	United Kingdom
USA	United States of America
USSR	Union of Soviet Socialist Republics

1 Introduction

Many of those who have been bedazzled by the great powers of modern science often lament that public communication of science through mass media in contemporary democracies has failed miserably to convey the values and method of science to the citizens. This failure, for them, has resulted from journalists' incompetence in understanding cutting-edge science, and their general disinterestedness in propagating science. Scientists often blame journalists for misrepresentation of their research in media.[1] There are complaints from the journalistic community also about the way scientists misguide them when approached for information.[2] However, it is always the journalists' lackadaisical attitude, from the dominant viewpoint, that shores up scientific illiteracy as well as an escalating regressive inclination towards superstitions, blind faith in religion and its demigods, leading to a complete eclipse of reason from social life. Some scholars have also noted a growing propensity among the media practitioners to 'sensationalize' and 'distort' science with the commercial interest of increasing readership/viewership and by extension the revenue from subscriptions and advertisements.

This understanding of science communication through mass media presupposes a positivist perception of science as a purely cognitive process through which the truth about physical reality is revealed. Scientists, from this vantage point, are the true custodians of knowledge. The conventional idea of science suggests that the scientific knowledge thus produced should be effectively utilized by society (in the form of technological applications) for social development and economic prosperity. Science communication to the general public hence is an important activity to transmit the advancements of science with an aim to inculcate a scientific bent of mind or 'scientific temper' in them for the betterment of everyday life. From this dominant point of view, popularization of science is, at its best, pedagogical, and, at its worst, 'distortion' of the genuine knowledge produced by the scientists (Hilgartner 1990). The positivist conception of science suggests that the general public, barring an elite minority of knowledgeable persons, is ignorant of science and its transformative powers, and hence need to be educated through science popularization. They are often stereotyped as a passive category although in certain contexts the public is portrayed as unruly, vehemently opposing scientific advancements which challenge their parochial worldviews.

More recent studies on science and media relationship move away from this conservative approach.[3] Dorothy Nelkin (1995b: 12) in her pioneering study points out that media is more than a communication channel between the scientific community and the general public. There is a complex process behind the production of science news, and 'public communication is shaped by cooperation and conflict among several communities, each operating in terms of its own needs, motivations, and constraints'. Hence she argues for more careful analysis of media presentation of science, for '[j]ournalists, their editors, and scientists themselves all influence the presentation of science in the press' (Nelkin 1995b: 13). It has been suggested since Nelkin's work that there is a steep and steady increase of science news in mass media and that scientists themselves are now more enthusiastic about positive news coverage of their research. 'Scientists [have] become more willing to talk, more prepared to comply with journalistic demands, and less squeamish regarding journalistic simplification, sensationalism, recontextualisation, and even inaccuracies in details', suggests a more recent study (Peters et al. 2008: 274).[4] Most of the scientific research institutions in the West have active public relations departments to facilitate interaction with media, thereby promoting their research and increasing public visibility. With a shortage of public funding for research and a consequent increase in cut-throat competition for resources, scientists often approach mass media for public attention that could reward them indirectly in securing funds (Nelkin 1995b; Weingart 1998). Top scientific journals like *Nature* and *Science* announce the about-to-be published research work in press conferences to increase citations to the paper and to boost the impact factor of the journal. The arrival of new media and social networking further complicates this picture. Researchers now resort to email groups, blogs, special websites, and Facebook pages to share and discuss their research with peer groups as well as a wider group of interested readers and the feedback helps improve the quality of research reporting to peer-reviewed journals (Weingart 1998). Sharing of preprints through internet repositories like arxiv.org and professional networking sites like researchgate.net helps scientists also to understand ongoing research in other laboratories in the field. There are also online discussion forums run by research communities as well as pirate websites for accessing otherwise copyright-protected resources. Scientists use the internet to get in touch with other research groups in distant parts of the world and to gather information about distant fields and disciplines. A similar use of popular science books, magazines, documentaries, and television and radio programmes by scientists to understand advancements in alien disciplines is well recorded (Bucchi 1998). When scientific controversies break out, it is observed that conventional communication channels (publication of research in journals, research exchange in seminars and conferences, and so on) collapse and researchers resort to popular communication channels like emails, websites, popular science magazines, and newspapers for cognitive resources, as Bruce Lewenstein (1995a) and Bart Simon (2001) have demonstrated in their case studies on cold fusion controversy in physics.[5]

On the other hand, mass media finds it important to report science more intensely due to growing public enthusiasm. Today, the public turns to mass media for scientific information and expert advice to deal with the risks they face in everyday life. This includes monitoring and safeguarding oneself from air and water pollution, exposure to carcinogens, risks of fast food and plastic, eating disorders, exposure to nuclear radiation, unknown side effects of medicines, reasons for the decrease in honey bee population in the locality, problems of living near a thermal power plant, and so on and so forth. Such risks are generated as a consequence of the modernization process guided by scientific and technological advancements of the modern world (Beck 1992). These risks are not easily predictable or manageable by science, and most of these risks are invisible but with lethal consequences. Often the public perception of risk comes in conflict with scientific assessments. Public controversies over science thus initiate conflicts in interpretation and a loss of trust in scientists. Mass media intervenes in such precarious situations and mobilizes expert advice from numerous sources and disciplines and knowledge traditions, and presents it for public scrutiny and deliberation. Attending more to the risk discourse and alerting the public about the effects of technoscientific advancements is rewarding to the mass media in terms of increased readership and circulation. Mass mediation of scientific advice to the public in the context of risks thus create new avenues for collaboration and conflicts between diverse social actors and institutions, often snowballing into big public controversies over risks.

Risk controversies[6] hence offer a unique opportunity to study the symbiosis between science and media in contemporary societies. A scientific controversy erupts when a dispute emerges among scientists or between scientists and laypeople on a scientific issue (Engelhardt and Caplan 1987:1). Controversies are often deliberated in different public forums and at particular junctures attain 'closure' instead of resolution, because of the difficulty to attain consensus over the issue under deliberation. Closure, as the term indicates, is a temporary, abrupt ending of the dispute. Although the scientific controversy disappears as a result of the closure, it may reappear any time and can have a second life (Engelhardt and Caplan 1987: 2; Nelkin 1995a: 456). Public controversies over science disclose the complex processes of scientific knowledge production and the otherwise hidden connections science has with diverse social institutions and actors. Scientific controversies allow us to examine science that is still in the making (Martin and Richards 1995: 513). The contingencies and ambivalences of 'science-in-the-making' are disclosed during such controversies (Michael 2002: 118). Similarly, Mendelsohn (1987: 99) opines that '[b]ecause it is in controversies that interests become most sharply drawn and the structures and processes of science most clearly tested, they provide a particularly strategic site for examining the complex set of relationships between the social and the conceptual in science. The conflicting claims of the contesting groups in a debate magnify (and potentially distort) the more commonplace daily procedures of scientific activity and thus give them greater visibility.' Hence a scientific controversy often makes public the internal tensions and ambivalences involved in scientific knowledge production, challenging the positivist notion of science as a disembodied, neutral, and objective way of truth seeking which is clearly separated from the domain of politics. Opposing the positivist apathy towards controversies as an 'unwanted externality' to scientific knowledge production (Mendelsohn 1987: 94), the sociology of scientific knowledge (SSK)[7] views scientific controversies as radically shaping modern science.

Public controversies over science turn out to be of special importance for mass media, as it invites journalists to face a completely different kind of science. During controversies, '[j]ournalists must cope with complex and uncertain technical information, conflicting scientific interpretations, and social biases that affect perceptions of risk' (Nelkin 1995b: 47). Unlike the journalistic reports on scientific spectacles (for example, reporting of a solar eclipse or of India's successful Mars mission in 2014 called Mangalyaan) and the reporting of a scientific discovery, science reporting during public controversies demands a completely different set of professional skills from media practitioners, in negotiating with the internal complexities of science as well as the contestations among the scientific community about the 'scientific truth' of the matter.[8] As the case study chapters show, science news is framed differently during controversies.

The research that culminated in the present book investigates the phenomenon of public deliberations of science in the context of emergent risk politics through a detailed analysis of three such controversies from Kerala being staged by the regional newspapers in early 2000. Kerala, one of the southernmost states of India, offers an interesting social context to look at mediated public deliberation of science in a non-Western context because of four main reasons. First, the region has a long history of domestication of science that goes back to the nineteenth century. The third magazine published in Malayalam[9] (*Paschimodayam*, 1847) was a popular science magazine. Modern science and the Enlightenment vision it embodied had deeply informed the public debates of the late colonial period.[10] Kerala has a long tradition of popular science reading. The rationalist discourse was prominent throughout the modern history of Kerala. Kerala Sasthra Sahithya Parishad (KSSP), the foremost people's science movement (PSM) of

India, has been active in the region since the early 1960s. Second, the developmental modernity of Kerala was enunciated quite significantly in a 'management oriented–scientific language' (Devika 2008: 181). The KSSP's interventions in the developmental discourse since the late 1970s epitomized the exceptional articulation of this language in developmental decision making. Third, Kerala witnessed a newspaper revolution in the 1960s itself, and even after the digital revolution of the 1990s and the ensuing growth of private television channels, newspapers continue to be the most authentic and efficient medium of political engagement for the ordinary citizens.[11] Kerala's vibrant political culture hence has the newspaper reading habit at its core (Menon 1994; Jeffrey 2000).[12] Finally, there is a steady increase in the number of public controversies over science appearing in the print media in Malayalam since the 1990s. These controversies often redefine politics itself in Kerala, as the case studies in this book demonstrate.

The key argument of this book is that media's role was central to the origin and development of risk controversies, and the close coupling of science and media in the context of risk politics led to the creation of a 'scientific public sphere'. Media had become the most prominent site of public engagement with science in Kerala by the 1990s, and the controversies analysed here represent the maturation of the scientific public sphere in the early years of the new millennium. This book contends that this new form of public engagement with science radically differs from its earlier form in the 1970s and 1980s nurtured by the KSSP. The present work discusses the social history of this shift and the resultant transformation in the characteristics of the scientific-citizen public of Kerala. Three of these controversies are analysed in this book to understand how risks were perceived, knowledge claims were contested, disciplinary rigidities collapsed, and trust in science and credibility of scientific institutions were re-negotiated. These public controversies demonstrate the formation of politico-epistemological alliances between different stakeholders and how scientists, journalists, editors, policy makers, government officials, and social activists encountered each other on the backstage to shape the deliberations on the front stage. Controversies also opened up the otherwise hidden dynamics of scientific knowledge production into full public view. Scientists' response to their overexposure to the media during the controversies as well as the interaction between researchers and journalists as two professional communities is analysed. Finally, the present study looks into how the public contestation of knowledge contributes to deepening democracy by re-instilling politics into science. Democratization of science under the agency of the scientific-citizen public, the book argues, was but constrained by its inability to include multiple publics and their alternative perceptions of risk in the political imagination.

Public Engagement with Science and Technology (PEST) in India

In India, research on public engagement with science and technology (PEST) is in its infancy as an academic field, although there is a growing scholarly enthusiasm in recent years. Many of the early studies were based on large, quantitative opinion surveys that assessed public attitudes towards science. Public attitudinal surveys were conducted in India by a small group of researchers from the National Institute of Science, Technology and Development Studies (NISTADS), New Delhi, in the 1990s. These surveys were inspired by the large-scale national surveys executed in developed nations as well as in Asian countries such as China and Japan to capture the level of scientific awareness (Raza, Dutt, and Singh 1996: 11).[13] A Science Communication Unit was constituted in the NISTADS in 1989 and the first large-scale survey to understand the level of public understanding of science (PUS) was carried out during the Kumbh Mela at Allahabad.[14] The team successfully conducted a series of surveys over the next few years.[15] These large-scale public opinion surveys were methodologically framed to move away from a dominant view that sought to find out percolation of scientific knowledge in society. Conventional surveys, according to the NISTADS group of researchers, failed to map the complexities of cultural reception and hence they proposed a more nuanced survey method that takes into account the 'cultural distance' between science and Indian public and the factors (such as class, rural/urban divide, education, gender, and so on)[16] that influence the same. They argue that

> [t]he hypothesis that the disciplines of science constitute a scale of proximity to the quotidian life of people where areas like astronomy and cosmology and geography and climate occupy the farthest end and

agriculture, health and hygiene could be placed at a distance that is quite close to their life, was reconfirmed [by the survey during Ardh Kumbh Mela in 1995]. The scientific information related to those areas which are at a shorter distance is likely to become a part of the people's structure of thinking without intense efforts but strategies for disseminating information related to other fields of knowledge have to be meticulously chalked out. (Raza, Dutt, and Singh 1996: 124)

These studies thus argued for a more complex analysis of public reception of science and hesitated to classify the public simply as 'scientifically literate' and 'illiterate' as the earlier opinion surveys did. At the same time, the NISTADS researchers continued with the deficit model of PUS that focused on assessing diffusion of scientific knowledge in society.

In more recent years, their methodological framework has become more nuanced (see Raza, Singh, and Dutt 2000, 2002, 2004; Raza, Singh, and Shukla 2009) to capture the socio-cultural and economic factors which structure the 'cultural distance between science and the people's cultural thought complex' (Raza, Singh, and Shukla 2009: 271). The idea of cultural distance is now defined as 'the distance that a worldview, attitude, perception, or an idea, generated within one cultural context, travels on a time scale for its democratisation within the thought structure of the other cultural sub-group(s)' (Raza, Singh, and Shukla 2009: 272). The diffusionist imagination of percolation of knowledge is intact here, wherein their approach is more interventionist to statistically gauge the cultural distance of scientific ideas in order to ensure a smooth flow of knowledge from the scientific community to the public. For instance, for the idea of 'solar eclipse' as a phenomenon that is deeply entrenched in non-scientific, traditional cultural belief system and associated with myths, religious belief and superstitions, the cultural distance is large (Raza, Singh, and Shukla 2009: 278). The complexity of a scientific idea and its abstract nature also may increase its cultural distance. Hence it is concluded from their quantitative surveys that 'various ideas face varying degrees of resistance' while travelling 'through the permeable membranes that define boundaries of a cultural grouping or a cultural sub-group' (Raza, Singh, and Shukla 2009: 278). Because of this diversity in cultural distance between science and different regional and local cultures, the strategies of public communication of science should be formulated to take into account the cultural distance so that science can be 'democratized'. The 'democratization' of science, from this perspective, implies an active intervention by the scientific community, science communicators, and policy makers to reduce the cultural distance. Democratization of science is thus an interventionist project of the scientific elite to infuse scientific ideas into the larger population to make science an integral part of the 'cognitive structure' of the cultural groups (Raza, Singh, and Shukla 2009: 278).

Another study that relied on public opinion survey to understand the public understanding of participation in the process of standardization of bottled water examined connections between public awareness of and trust in standards, as well as public perception of their participation in the standard setting exercises (Bhaduri and Sharma 2014). As part of the study 97 respondents were interviewed at various locations in New Delhi and the researchers reached the same conclusions as of the NISTADS group: cultural distance was influenced by a variety of socio-economic and cultural factors such as education, exposure to mass media, and urban/rural divide. Although arguing for 'a framework beyond the simplistic deficit–dialogue dichotomy', the study reverts to the quantitative methods employed by the proponents of the deficit model and fails to closely examine the nuances of the public controversy over the standardization process of drinking water in India (Bhaduri and Sharma 2014: 484).[17]

A major problem with the diffusionist-interventionist framework is that the production of scientific knowledge is seen as a socially disembodied, esoteric activity which is exclusively under the epistemological authority of scientists. The public, from the purview of the framework, is deficient of scientific knowledge, and hence is to be educated through various forms and methods of public communication of science. Against this positivist perception of science, a more sociologically informed understanding of modern science as a 'diffuse collection of institutions, areas of specialised knowledge and theoretical interpretations whose forms and boundaries are open to negotiation with other social institutions and forms of knowledge'[18] is required to explore how publics engage with science in a variety of micro-sociological contexts which the large-scale opinion surveys fail to capture.

Biswanath Dash's study that examines villagers' engagement with the cyclone warning mechanism of the India Meteorological

Department (IMD) in the coastal regions of Odisha (2014a, 2014b) is a significant attempt to capture this dynamics.[19] His research was based on ethnographic exploration in 22 cyclone-affected villages of Odisha in three phases over three years (2007–9), during the months of the pre-monsoon cyclone season (October–November). The study found that the IMD's expert advice was of little significance for the rural public because of the failure of the formal cyclone warning mechanism to take into consideration a set of local geographical factors and social realities that were excluded from the warnings but which the villagers found central to their decision regarding staying or evacuating the locality (Dash 2014a). The villagers were not antagonistic to the cyclone warnings issued by the IMD, but they did not blindly accept them either, and their own local knowledge about the climatic conditions heavily informed them in their decision-making process (Dash 2014a). Dash's study employed insights from the contextualist (or constructivist/ dialogue) model of PEST and the social representation theory (SRT). While the latter helped to argue that different systems of thinking and knowing coexist and fulfil different purposes and functions in social life, the contextualist model helped in examining the social and cultural context of public engagement with science. Unlike the earlier studies which were informed by the deficit model, the researcher employed qualitative methods to capture the villagers' encounter with scientists in micro-sociological contexts, yielding a completely different picture of PEST, where the public was knowledgeable and actively engaging with science. This approach also avoids the interventionist accent of the previous studies, and suggests the need for 'institutional reflexivity' of science to engage with publics and their 'lay-knowledge' to be more socially relevant, as Brian Wynne has suggested in his classic contextualist study of the Cumbrian sheep farmers' encounter with scientists (see Wynne 1996).[20]

The present study explores an important aspect of PEST—the role of media in the process of engagement. Bringing media into the picture opens up novel possibilities to theorize the public beyond the concept of 'mini-publics' tied to micro-sociological contexts as proposed by the contextualist model: rather, 'the mediated publics of science' opens up new possibilities to explore the dynamics of public engagement with science in liberal democracies which are deeply shaped by the presence of mass media. While micro-sociological research identifies media as a social actor in the contexts of engagement,[21] the present study theorizes its pivotal role in shaping a mediated public that engages with science in the 'scientific public sphere' of Kerala.

Although media often stages public controversies over science, this function of the media has not been sufficiently addressed in the Indian context. There are some studies which examined journalistic production of science news with an objective to quantify science news to see how scantly science is reported in Indian dailies. These studies examined the role of media in communicating science to the public from a deficit-interventionist point of view, where the emphasis lies on how to improve science reporting by identifying the hindrances in the smooth flow of information through mass media to the reader-public. Pioneering efforts in this direction were done in the late 1980s and early 1990s. Interestingly, one of the earliest studies was on the regional newspapers of Kerala. J.V. Vilanilam's study was conducted during 1987–8 on ten dailies (eight Malayalam and two English, which had circulation in Kerala) and eight magazines (all Malayalam) to understand the news space allotted to science and technology (S&T), the sub-categories of these news reports and their frequency, the reasons behind the dominance of certain sub-categories of science news in newspapers, and the ways to improve S&T reporting in mass media (Vilanilam 1993: 179–80). All issues of the newspapers and magazines during the period of study were examined. The research suggested that S&T news in both English and Malayalam dailies and magazines were insufficient. Among the newspapers analysed, English dailies outdid their Malayalam counterparts: *The Hindu* devoted maximum space for S&T, followed by *The Indian Express* (Vilanilam 1993: 182). Among the Malayalam newspapers, *Deshabhimani* devoted maximum space for S&T (Vilanilam 1993: 182). Two sub-categories of science news which had prominence were agriculture and agricultural sciences, and public health, medicine, and medical technology. Media owners' increased focus on enhancing their business and industrial interests as well as an over-dependence of the mass media on advertisements for revenue, according to the researcher, are the reasons behind a negligible interest of mass media in reporting S&T-related news (Vilanilam 1993: 188). The researcher suggested that grassroots organizations like the KSSP have a central role in science popularization, and '[w]hen such organisations empower people, the big media will automatically change their contents and style of operation' (Vilanilam 1993: 188).

The large-scale opinion survey among the residents of Mangolpuri in New Delhi conducted by a team of researchers from the NISATDS in 1994 looked for the main channels of scientific information about plague, and the accessibility and efficacy of mass media as an important channel of communication (Raza, Dutt, and Singh 1996: 33). Through field survey on media effect, the researchers found that government-owned television was more effective in terms of accessibility and reliability in delivering scientific communication than radio, newspapers, and interpersonal interactions in Mangolpuri (Raza, Dutt, and Singh 1996: 33–4). Following this opinion survey on media effectiveness in communicating scientific knowledge to the city dwellers in the context of a plague epidemic, the researchers also analysed a sample of news reports from the English national dailies 'to quantify the coverage [of plague-related science news] in terms of number of items under various categories constituted, the space allocated by the media to various plague-related stories, and their column spreads' (Raza, Dutt, and Singh 1996: 76), an attempt similar to the study by Vilanilam (1993). Their analysis showed that the role of media was largely positive in the dissemination of scientific knowledge. In another nation-wide study carried out by the NISATDS team, news reports on S&T in 27 English newspapers published in different parts of India were analysed during 1996 (Dutt and Garg 2000). They used the technique of counting columns to quantify the amount of space allocated for science news by each newspaper. They found that nuclear S&T was the largest reported subject (16.67 per cent), followed by defence and space (16.46 per cent), and astronomy (15.44 per cent). It was *The Pioneer* that devoted maximum space for S&T news followed by *The Hindu* and *The Times of India* (Dutt and Garg 2000: 127–8). They also analysed the sources of news ('Indian' or 'foreign') and the importance given to S&T-related news (whether the report appeared in the front page or inside). They concluded that 'far less than one per cent of the total printed area was occupied by the science and technology items' and complained that '[e]ven this low coverage was not regular and sustained' and '[o]nly those [S&T-related] stories that had political implications found the slot on the front page while the majority was placed on the inside pages' (Dutt and Garg 2000: 138). Many studies of a similar vein used quantitative content analysis to gauge the effectiveness of mass media in science communication, and recommended strong intervention from the state, policy makers, and scientific community to ensure better flow of scientific information to the media audience.[22] The significant role of governmental agencies like Vigyan Prasar in facilitating better communication by organizing training workshops for journalists and by providing popular science resources to mass media institutions was highlighted.[23]

There are a small number of studies on science reporting in Indian mass media which employ narrative analysis. Bharadwaj (2000) examines the 'narrativisation of infertility' by studying selected texts from English newspapers about assisted conception in India. The author, through narrative analysis of the texts, suggests a close alliance between medical practitioners and journalists that leads to favourable reporting of assisted reproduction in the newspapers. Experts have successfully mobilized media support through their backstage collaboration with journalists to acquire public credibility (Bharadwaj 2000: 70). The researcher contends that 'the media/medicine symbiosis' has successfully hid from public view the politically constructed natures of the technological structure and generated a demand for these technologies and their experts through the media narratives (Bharadwaj 2000: 76). The study suggests that science–media collaboration is largely detrimental to public interest and reveals how the political dynamics of medical technologies is being black-boxed from the public. Pillai (2012) has a more nuanced take on the role of the mass media in contemporary risk societies. The author argues,

> In risk societies the media capitalises on the insecurities and fears of the reading/viewing public and takes upon itself the self-professed role of moulding a critical public and alerting this public to the notion of risk. Media becomes the public eye through which risk is perceived, and the politics of both magnifying the risk as also its denial are staged by and through media. (Pillai 2012: 32)

In the context of a controversy over an old dam in the Mullaperiyar region, she examines the role of media in Kerala and Tamil Nadu to see how risk is differentially constructed through media discourse. However, her approach to science is externalist like the previous study on assisted conception technology; both the authors fail to investigate how contestations over S&T are mediated, and largely end up demonstrating how scientific and technological narratives are employed by

media to catalyse fear and anxiety about the risks involved. How risk
discourse has transformed science, media, and politics, however, largely
remains outside the analytical radar of these studies. Similarly, both
the studies decontextualize the news-texts they analyse by selecting a
few English news reports as representative texts while neglecting a vast
number of news reports generated by the regional language press with
regard to the controversy.[24] As we shall see, the regional language newspapers offer a much nuanced dynamics of the science–media coupling
and the construction of mediated publics.[25]

Three public controversies over science that were widely deliberated
in the Malayalam newspapers in the beginning of the new millennium
are empirically examined in the book.[26] These controversies are selected
for their sustained reporting in the press and their strong existence in
public memory even today. Also, the three controversies analysed here
surfaced at the turn of the millennium, and they simultaneously epitomized a shift in the public's engagement with science in Kerala, which
is discussed at length in the book. The first case is the controversy that
erupted in the wake of certain clinical trials conducted at the Regional
Cancer Centre (RCC) at Thiruvananthapuram, the state capital of
Kerala. The research team led by the director of the institute administered two derivatives of a chemical to the cancer patients admitted in the
institute, allegedly without their prior consent. The RCC research team
carried out these clinical trials as part of a research project of the Johns
Hopkins University (JHU) that aimed at developing a new drug for oral
cancer. The matter leaked into the public domain when some researchers from the institution protested the non-employment of appropriate
scientific procedures during the drug trial. The issue snowballed into a
major scientific controversy staged by the regional press that witnessed
participation of a large number of actors in the deliberation. The RCC
clinical trials controversy raised serious questions regarding unethical
practices in scientific research and the internal tensions and contestations for power within the scientific community were made public.

The second case study is a scientific controversy that erupted in
the wake of a series of tremors that struck Kerala in 2000–1. Different
groups of scientists provided diverse and often contradictory scientific
explanations for an array of geophysical phenomena (such as well collapses, tunnel formation, and unseasonal leaf-fall) reported from all
over the region, resulting in a strong public debate on the legitimation

of scientific expertise.[27] The regional press revealed the internal ambivalences of the knowledge production process while providing a deliberative space for a wide range of actors including mutually contending groups of scientists from different institutional and disciplinary backgrounds. The case suggests that the media deliberation in the course of a scientific controversy offers valuable insights into the public's participation in the deliberative process concerning the science of the phenomena.

The phenomenon of 'coloured rain' that began to appear towards the tail end of this controversy is the third case examined. The spread of coloured rain in the region triggered off a big controversy in the regional press when the explanation provided by the scientists at the Department of Physics and Meteorology of the Centre for Earth Science Studies (CESS) was challenged by experts from other disciplines and institutions. Gradually a group of scientists who were working on a less scientifically established theory of origin of life (theory of Panspermia) successfully 'explained' the coloured rain phenomenon as evidence for extra-terrestrial life with the help of media, creating an epistemological crisis for the regional/international scientific community. Following this the public deliberation entered an unexpected terrain wherein the boundaries between science and non-science were constantly being remapped.

News reports in five major Malayalam newspapers are analysed here—*Malayala Manorama*,[28] *Mathrubhumi*,[29] *Deshabhimani*,[30] *Madhyamam*,[31] and *Kerala Kaumudi*.[32] The first two among them (*Malayala Manorama* and the *Mathrubhumi*) shared 87 per cent of the total readership in Kerala in the late 1990s (Jeffrey 2000) and the other three newspapers held the third, fourth, and fifth positions respectively at the time of the unfolding of the controversies under analysis. Content analysis of these newspapers provides primary data for the case studies; the news reports, special science articles, editorials, letters to the editor, and visual images (photographs, cartoons, diagrams, and so on) are examined sequentially from the beginning of each controversy up to its closure. Daily reporting of the controversies in five newspapers were followed,[33] and the discourse patterns gradually emerged were closely observed. This was important because each day's reporting in different newspapers cumulatively influenced the news presentation on the next day. In other words, each journalist/editorial team puts effort to understand how the controversy was reported in other newspapers on the same day, and this has influenced his/her (or the newspaper's) news presentation the next day. Actors' responses and reflections on the reported events and issues on the next day's newsprint further oriented the controversy. And this process continued throughout the unravelling of the public debate. Therefore, the method of content analysis was not dependent on a select set of news texts; the approach was more ethnographic in nature: the emphasis was on the gradual evolution of the controversy through the interactions and performances of the actors at the front stage of the scientific public sphere on an everyday basis. Each news report about the controversy thus became important for analysis. That is to say, I have collected and analysed almost every single news report connected with the controversies under study. The case study chapters thus provide a thick description of the controversies based on the news analysis that followed the utterances and performances of actors on the front stage on a day-to-day basis. Technical information and scientific terms and categories are used in the book as they developed in the course of deliberation in the scientific public sphere, and hence not necessarily conforming to any 'standard' scientific lexicon. The discourse analysis of media deliberations is supplemented by detailed semi-structured interviews with the scientists, journalists, and social activists who had participated in the scientific public sphere (either as a backstage manager, or an actor on the front stage). This strategy has helped to understand the backstage management of the controversies.

Organization of the Book

The next chapter of the book theorizes the science and media coupling by engaging with relevant literature on the subject. Different approaches to the problem have been critically assessed to conceptualize the scientific public sphere and the mediated scientific-citizen public. The chapter argues that scientific public sphere is methodologically helpful in studying the historical evolution of the scientific-citizen public and the emergence of mass media as the prime site of public engagement with science in recent times. The third chapter introduces the scientific-citizen public of Kerala. The regional press, due to historical reasons, became the prime site for staging deliberations on science during the risk controversies in the 1990s. This has generated a scientific public sphere in the region, a separate deliberative space for debating science. By tracing these transformations in the historical context of Kerala, the chapter argues that the scientific-citizen public who engaged with science during the 1970s and 1980s became reconfigured into a mediated public in the 1990s onwards, in the wake of intense manifestation of risks in everyday life of the citizens.

The next three chapters present the scientific controversies that help in understanding the functioning of the scientific public sphere in Kerala. These controversies highlight the central role the regional Malayalam press has assumed in recent decades in mobilizing a deliberative space to debate science. The fourth chapter discusses how a scientific controversy on clinical trials that started as an internal conflict between two groups of researchers within the RCC triggered off a public deliberation on the regulation of scientific research and redesigning of ethical protocols for drug research. The credibility of the cancer research institute was at stake, and the public trust in the researchers deteriorated. The chapter demonstrates how the trust in scientific community is regained and the credibility of the institute restored through a complex deliberative process wherein media institutions, scientists, health activists, government officials, and national regulatory bodies participated.

The fifth chapter presents the contestation over knowledge claims by different groups of experts—when a series of micro-earthquakes struck Kerala, various scientific institutions provided contradicting explanations regarding the epicentre and magnitude, leading to public criticism of scientific expertise. The shortcomings of scientists to offer a satisfactory explanation to the series of unnatural geophysical phenomena that appeared following the tremors opened up the internal dynamics of scientific knowledge production to full public view. Newspapers sustained the controversy by bringing forth the contradictions in expert opinion and highlighting the public disbelief in the scientific explanation against the possibilities of any major earthquake in Kerala. The debate became focused on the question of reservoir-induced seismicity as the epicentre of the quakes was near the Idukki dam, on which most of the electricity generation of the state depended upon. The gradual shift of deliberative focus from the potential threat posed by the Idukki dam to the old Mullaperiyar dam (used for irrigation in Tamil Nadu) is analysed to expose how regionalism and inter-state conflicts played a crucial role in bringing closure to the controversy. Following this, a series of bizarre geophysical incidents were reported from the region, triggering off a new spate of controversy. Well collapses began to be reported from the nook and cranny of Kerala, and media played a central role in prodding the scientific community to do in-depth research on these phenomena. Not satisfied by the official scientific brief from the CESS, Thiruvananthapuram, the media actively supported alternative scientific interpretations and studies.

How do scientists mobilize media resources in the context of a public controversy to accrue scientific credibility to their research is the focus of the sixth chapter. When the phenomenon of coloured rain struck across Kerala in 2001 followed by a thunderous meteoric explosion in central Travancore, a group of researchers who were working on the theory of Panspermia proposed that the red colour of the rain was due to biological cells of meteoric origin, against the biologists' argument that the reason was algal and fungal spores. Skilfully promoting their findings on red rain and introducing their less-scientifically legitimized field of research to the public with the help of global media, these researchers challenged the epistemological, disciplinary, and institutional hierarchies of science. Research communication in the form of preprint with the help of online portals and conference proceedings along with their collaboration with the Cardiff Centre for Astrobiology (UK) were pivotal in their path to fame. The chapter discusses the renegotiation of the boundaries between science and non-science and conflicts over public legitimization of a theory versus its non-validation by the regional scientific community.

The seventh chapter is on journalistic production of science news. The backstage dynamics of the scientific public sphere is examined here. The rhetorical strategies employed in the presentation of science news as well as the mobilization of resources for news production by the media are discussed. The skilful deployment of news genres such as editorials and letters to the editors as well as the framing strategies are examined to demonstrate how a public controversy is actively produced and maintained by the media. The chapter argues that science communication through regional press is a complex professional activity where scientists and journalists actively participate and associate with each other for mutual professional benefits.

Insights from the previous chapters on the coupling of science and media are linked in the concluding chapter to the theoretical debates on the 'deliberative turn' in public engagement with science. The deliberative turn suggests that a more inclusive, wider, and intense public deliberation of science is essential for greater democratization of science, and acknowledges the central role of mass media in it (Jasanoff 2003b; Wynne 2003). The chapter suggests that the science–media interaction in the context of risk controversies in Kerala produces a particular kind of democratic deliberation, and its boundaries are set by the scientific-citizen public. Characteristics of this mode of deliberation are highlighted based on empirical analysis done in the previous chapters. The chapter concludes with observations on the transformations in the mode of public engagement with science in Kerala and its impact on democratization of science.

Notes

1. 'Media' and 'mass media' are used interchangeably in the book.

2. A recent example for such views is the two-day seminar titled 'When Science Meets the Public: Bridging the Gap' held at the National Institute of Advanced Studies (NIAS), Bangalore, where Srikumar Banerjee, former chairman of India's Atomic Energy Commission, and Deepak Pental, former Vice-Chancellor of the University of Delhi and a leading geneticist, strongly criticized the way genetic engineering and nuclear technologies were reported in media. Scientists who attended the seminar expressed their lack of confidence that journalists could 'write factually accurate, and at the same time interesting articles' since the journalists 'are out to sensationalise events and sell newspapers or increase television viewership, as the case may be.' On the other hand, journalists pointed out 'problems in accessing experts and information' and complained that '[i]nstitutes do not update the media about their research, and their efforts at communication are usually reactionary, in response to a crisis situation'. See the report of the seminar, 'Scientists and Journalists Call for Better Communication', *Asian Scientist Magazine*, 2 July 2014 (http://www.asianscientist. com/2014/07/features/scientists-journalists-communication-2014/, accessed on 30 March 2016).

3. See Secko, Amend, and Friday (2013) for an elaboration on this recent shift from conventional pedagogical models to more nuanced participatory models of science communication. For a brief introduction to the field of science and media, see Lewenstein (1995b).

4. The study was based on data gathered from researchers in France, Germany, Great Britain, Japan, and the United States.

5. See the case studies in Rödder, Franzen, and Weingart (2012) for many of these aspects of science–media interaction noted here.

6. Public controversies over science, risk controversies, and scientific controversies are used interchangeably in the book.

7. SSK emerged in the late 1970s as a departure from the earlier sociology of science that emphasized the sociological characteristics of the scientific community and their social organization. Robert K. Merton was the most prominent figure of this conventional trend. Against this, and under the influence of Thomas Kuhn's revolutionary work on scientific knowledge production that challenged Karl Popper's propositions on scientific method (Kuhn 1996), the 'Strong Programme' of the Edinburgh School in SSK argued that science is not simply influenced externally by the social, but its cognitive core itself is socially constructed. Similarly, ethnographic studies on science (laboratory studies) examined how scientists do science on an everyday basis in laboratories and fields and demonstrated that scientific knowledge production is like any other everyday social activity. See for Strong Programme, Bloor (1991), and for laboratory studies, Collins (1981), Knorr-Cetina (1981), Latour (1987), and Latour and Woolgar (1979). For a general introduction to science, technology, and society (STS) studies, see Jasanoff et al. (1995) and Sismondo (2004).

8. One of the earliest studies in this regard from the purview of SSK is Collins (1987).

9. Malayalam is the major language spoken in Kerala.

10. For example, the debates on the theory of evolution and scientific rationality in Indulekha (1889), one of the earliest novels in Malayalam.

11. See Mehta (2008) for the growth of Indian satellite television in the 1990s. India has the second largest newspaper industry in the world (after China) with 72 million copies being sold every day. Newspaper holds the status of the most credible source of news in India even today (Sahay 2006: xvii–xviii).

12. A significant image of the newspaper-reading culture in Kerala is the sight of people reading and discussing the day's news in public places in the morning. The beedi workers in northern Kerala used to listen to the news of the day during work as a fellow labourer read out newspapers for them. See Isaac, Franke, and Raghavan (1998: 40). However, the centrality of newspapers in constituting the political culture of the region seems to be changing in recent times with the mushrooming of private news channels. See Venkiteshwaran (2014).

13. Two major public opinion surveys in the West today are the National Science Foundation (NSF) survey in the United States and the Eurobarometer survey in Europe.

14. The study findings were published as Raza, Dutt, and Singh (1991).

15. These surveys included the one conducted during the Ardh Kumbh Mela at Allahabad in February 1995 that interviewed 2,700 people (Raza et al. 1996); an urban survey at Mangolpuri, Delhi, in 1991 (14,600 respondents)

(Raza et al. n.d.); and another survey at Mangolpuri during a plague outbreak in Maharashtra and Gujarat in 1994 (Raza, Dutt, and Singh 1996, 1997).

16. For instance, Raza et al. (n.d.) conclude that students and white collar employees scored high in their knowledge of science during the survey in Mangolpuri (1991). They suggest that '[t]his also confirms that to absorb and understand increasingly complex phenomena socialisation in the tradition of modern science and working environment play an important role'. Similarly, migrants from rural areas scored better than the urban dwellers on questions regarding agriculture (Raza et al. n.d.).

17. The public controversy was in the aftermath of the declaration of the results of a laboratory assay of the samples of bottled drinking water sold by major brands in India in 2003 by an environmental non-governmental organization (NGO) called Centre for Science and Environment (CSE) based in New Delhi. See Bhaduri and Sharma (2014).

18. See the editors' introduction in Irwin and Wynne (1996: 8).

19. My own research on various dimensions of PEST in India critically engages with the contextualist model in a variety of contexts. See Varughese (2008, 2011, 2012, 2014).

20. See the next chapter for a detailed discussion on Brian Wynne's study. For one of the earliest discussions on the contextualist model in India, see Raina (1993, 1999).

21. For example, Addlakha (2001) observes, in the context of Dengue fever outbreak in Delhi, that the media plays multiple roles as commentator, communicator, educator, and watchdog, linking the medical profession, state, and local communities.

22. See the studies on mass media that appear in the *Indian Journal of Science Communication* and the *Science Communicator*. See also Salwi (2002), Mazzonetto (2005), and chapters in Chakraborty (2001) and Patairiya and Nogueira (2011).

23. Vigyan Prasar is an autonomous organization for public communication of science under the Department of Science and Technology (DST), Government of India, established in 1989, but ventured into large-scale science popularization programmes only since 1994. It provides ready-to-print science pages to Indian newspapers with an objective to help Indian print media to increase science coverage. It also prepares science related programmes for television and radio. Vigyan Prasar organizes training programmes for journalists and science writers. For more information on its activities, see Padmanabhan (2004).

24. See also Nambiar (2014) and Saeed (2013) for similar analyses of how media construct environmental and developmental discourses in India.

25. See Rajagopal (2001) for the cultural dynamics of vernacular media in opposition to the elitist English media, a cultural phenomenon that creates a 'split public' in India.

26. What can possibly be called a pioneering investigation into scientific controversies in Kerala has been conducted by Nair (1999). The author investigates the scientific controversy that emerged in Kerala in 1994 regarding the alleged 'spy work' that involved a scientist from the Indian Space Research Organisation (ISRO) and two women from the Maldives. However, the book falls into the category of investigative journalism rather than being academic in orientation. The author blames the media (especially the role of *Kerala Kaumudi* and *Deshabhimani*) and politicians for forging the controversy against the scientist/institution.

27. There were 28 kinds of geophysical phenomena reported in media. See Appendix.

28. The newspaper was launched on 14 March 1888 by Kandathil Varghese Mapilla as a weekly publication from Kottayam. It has grown into the biggest newspaper company in Kerala over the twentieth century.

29. *Mathrubhumi*, a daily, was launched on 18 March 1923 from Kozhikode by K.P. Kesavamenon with an objective to support the national movement in Kerala.

30. *Deshabhimani*, the organ of the Communist Party (now CPI[M]) of Kerala, started publishing as a weekly magazine from Kozhikode in 1942.

31. *Madhyamam* was established in Kozhikode in 1987, and is published by the Ideal Publications Trust under the patronage of the Jama-at-e-Islami of Kerala. According to the website of the newspaper, '[i]t is the largest circulated Asian Daily in the Gulf, and the first International Indian Daily' (http://www.madhyamam.com/aboutus/history.html, accessed on 12 March 2016).

32. *Kerala Kaumudi* was launched in 1911 from Thiruvananthapuram by C.V. Kunjiraman, with his son K. Sukumaran as the founder editor. The newspaper has wide readership in southern Kerala.

33. The Thiruvananthapuram editions of these newspapers were used for the content analysis.

2 Science, Media, Risk Politics
Constructing a Scientific Public Sphere

Science, with its positivist self-image, claims a special epistemological status in society. This self-image of science gradually developed along with its historical transformation from the seventeenth century 'gentlemen science' to the disciplinary differentiation in the nineteenth century. By the mid-twentieth century, it completed its transformation into 'big science', a complex professional enterprise that requires large-scale financial and infrastructural investments as well as collective intellectual collaboration and strong internal quality-assessment mechanism, being part of the military–industrial complex.[1] Science was taken under the wings by the nation-states and it continues to be an integral part of the liberal democratic imagination.[2] The nation-state's social contract with science assured the latter with political patronage but non-intervention in its internal organization. The scientific community was granted autonomy to govern knowledge production and in return the nation-state benefitted from their intellectual output that was harnessed for national prosperity and progress (Guston 1992: 2). A clear separation is thus established between truth and power; science has been the custodian of truth about natural reality, and politics, the domain of power over the social. Science was granted autonomy in producing truth, unaffected by politics, and in return the state has actively sought technical solutions from science to solve societal problems. Actually, this reveals a paradox of liberal democracy; giving special status to an elite social group (scientific experts) subverts the very democratic principle of equal representation (Turner 2001; Fischer 2005). Expert opinion thus ironically became fundamental to the smooth functioning of liberal democracy. The distinction between truth and power, and the technical and the political thus clearly placed science hierarchically above politics (Ezrahi 1995).[3]

Such a peculiar relationship between science and politics constantly demanded citizens' persistent appreciation of science. Science popularization programmes under the auspices of the nation-states aimed at greater public appreciation of science for its epistemological capabilities in providing solutions to the problems of liberal democracies. Public communication of science through mass media was perceived as integral to ensuring greater public appreciation of science. A professional devotion from the media was demanded to serve science by faithfully transferring scientific knowledge to the public, thereby annulling any possibility for constituting a vibrant public sphere that critically engaged with science.

However, these convictions that shaped the social contract of science in contemporary democracies soon began to dissolve. It has been noticed that there is an intensifying 'loss of distance' between science and politics as well as science and media, leading to a series of changes in the internal differentiation and structural linkages of science with its system environment since the 1960s.[4] Such a loss of distance jeopardized the demarcation between truth and power. This chapter delves into conceptualizing these transformations observed in science's relationship with other function systems and its impact on shaping critical publics in today's liberal democracies. The key argument here is that the loss of distance of science vis-à-vis media and politics has generated a *scientific public sphere* that gets activated through public controversies over science, producing a scientific-citizen public that determines the nature of public engagement with science in contemporary risk societies.

The Changing Social Contract of Science

The social contract of science that epitomized the accommodation of 'the modern political constitution of truth that separated the political and scientific realms' has gradually begun to transform and a new relationship is emerging around the increased demand for public accountability of science (Guston 1992: 15). In the 1960s, the relationship between science and politics began to be restructured with the growing uncertainties and complexities in managing risks. The emergence of a risk society in the late modern, post-industrial world also gave rise to a growing public scepticism on science's exclusive capabilities to manage risks (Beck 1992). Creation of risks has increasingly been perceived as a consequence of the process of modernization that was closely coupled with science and technology (S&T) in the last two centuries. Because of this relationship between the modernization process and S&T, risks cannot be managed by scientific experts alone (Beck 1992). Risks are no longer tangible and exclusively manageable by science. The proliferation of risks has thus created a crisis for liberal democracy: now science cannot provide technical solutions to social problems. Many of these risks (for example, nuclear radiation or toxic pollutants) were not 'visible' without 'the "sensory organs" of science—theories, experiments, measuring instruments' (Beck 1992: 27).[5] At the same time, the scientific interpretation of risk has been deeply in conflict with the risk perception of people in their everyday life, which has created a split between technical and social rationalities and has realigned them in a more democratic relationship with each other. According to Ulrich Beck,

> What scientists call 'latent side effects' and 'unproven connections' are for them [the people] their 'coughing children' who turn blue in foggy weather and gasp for air, with a rattle in their throat. On their side of the fence, 'side effects' have *voices, faces, eyes and tears*. And yet they must soon learn that their own statements and experiences are worth nothing so long as they collide with the established scientific naiveté. The farmers' cows can turn yellow next to the newly built chemical factory, but until that is 'scientifically proven' it is not questioned. (Beck 1992: 61)[6]

The split between technical and social perceptions of risk leads to the citizens' search for their own cultivation of lay-expertise and interpretations of the ontology of risks they face in everyday life, which often conflict with the scientific assessment (Beck 1992). The crisis is that 'the sciences are *entirely incapable* of reacting adequately to ... risks, since they are prominently involved in the origin and growth of those very risks' (Beck 1992: 59).[7] The plurality in risk perception and the consequent conflicts between experts and the lay-public led to the burgeoning of a 'subpolitics' (of science) that destabilized the social contract of science founded upon a clear distinction between the domains of science and politics and the functional relationship between them (Beck 1992). This subpolitics brought science back into the domain of politics;[8] the technical is now subsumed into the political unlike in the previous phase. This obviously has a tremendous and long-lasting impact on the political system—politics is no more dependent on technical advice from experts. Rather, technical advice itself is now a subject of political contention. The new advancements in biosciences and nanosciences challenged the dualism of nature and society, adding to the risk politics and the public demand for more public accountability from experts (Jasanoff 2003a, 2003c). Because of these transformations in relation with the risks resulting from modernization, now 'natural scientists work in a *powerful political, economic and cultural magnetic field*' due to 'the obviously political character of their "subject"', suggests Beck (1992: 82).[9]

Beck's theorization of the risk society is helpful in understanding an emerging process of reflexive modernization that has also witnessed a change within the knowledge production practices, and science's willingness to be publically accountable and engaging with a wider public. However, this process of structural reconfiguration of science is hardly of interest to the risk society framework, apart from references to science becoming reflexive about its own role in generating/ managing risks. Those scholars in science, technology, and society (STS) studies who examined the changing configuration of science in the neoliberal context in late modern/post-industrial societies have framed these transformations either as the emergence of a new mode of knowledge production (mode-2 or post-academic science) that coexists with conventional science[10] or as creating new forms of knowledge oriented towards technical governance of risks such as regulatory science (Jasanoff 1987, 1990) or post-normal science (Funtowicz and Ravetz 1992). Proponents of mode-2 knowledge have noted a growing contextualization of science; science has become a more open system in the context of a 'growth of complexity in society', leading to high manifestation of uncertainty in both (Nowotny, Scott, and Gibbons 2001: 4–5). The salient features of mode-2 knowledge, for them, are its production within the context of application, transdiciplinarity, multiplication of the sites of knowledge production, high self-reflexivity, and new forms of quality control (Nowotny, Scott, and Gibbons 2003: 186–7). All these five characteristics are new to the system of science (Nowotny, Scott, and Gibbons 2003). The 'regulatory science' (also known as 'policy relevant science'), on the contrary, is the differentiation of a new mechanism at the intersection of science and politics, 'produced to support governmental efforts to guard against risk' (Jasanoff 2003c: 229). Regulatory science involves the service of experts from diverse disciplinary backgrounds (including social science disciplines) in providing technical advice on risks (Nowotny, Scott, and Gibbons 2003: 233). Similarly, Funtowicz and Ravetz (1992: 750) propose the emergence of a 'post-normal science' during 'high decision stakes and high system uncertainties' parallel to 'normal science' and 'consultancy science' which devices an 'extended peer review' by a spectrum of stakeholders.[11] Both these theorizations about the functional differentiations within contemporary science in dealing with risks seek for the creation of more socially inclusive and publicly accountable methods of decision making. Mode-2 science is also expected to acknowledge and work through socially distributed expertises and multiple sites of knowledge production that include civil society spaces. This indicates that science is going through a rigorous internal differentiation which has not completely manifested yet. Nonetheless, in its attempt to govern risks, the new science aligns itself with democratic politics.

The pertinent question is regarding the impact of these changes on the S&T system of developing countries like India. What is the impact of the changing social contract between science and society on democratic politics outside the West? How do the public(s) of science get shaped in the world risk society? How does the reconfiguration of science impact the public spheres of contemporary democracies? What is the nature of the multiple publics emerging at the intersection of science and politics in the globalized world? What is the role of mass media in this new mode of public engagement with science? What is the social dynamics of the encounter of the mediated publics with science? The present study attempts to address these questions from a South Asian context. The rest of the chapter develops scientific public sphere as a theoretical construct to embark on such an exploration.

The 'Deliberative Turn' in Public Engagement with Science and Technology (PEST)

Although the risk society thesis refers to a public that is produced through and engaging in the risk discourse, it does not attempt to theorize it adequately, as its energy is spent more on the characteristics of the emergent risk society. It is the contextualist model (also known as the constructivist model) within the PEST emerging in the 1990s that attempted to do so, by engaging with the question of risk.[12] The model's attempt was to conceptualize specific, located publics who actively engage with scientists in micro-sociological contexts. The contextualist model challenged the conventional (and also dominant) diffusionist model by emphasizing the active role of the local communities and citizen groups in negotiating the meanings of science in their daily life. Studies from the contextualist framework suggested that the citizen groups/local communities possess lay-knowledge/expertise based on their day-to-day experiences and that scientists should therefore engage with these publics to benefit from them in designing research in better, socially robust ways. The model attempted 'to investigate how people experience and define "science" in social life, and how particular scientific constructions incorporate tacit, closed models of social relationships that are or should be open to negotiation' (Wynne 1995: 362). Empirical studies from this perspective revealed the role of science in 'framing' public debate and the implicit social framing of science itself (Irwin and Wynne 1996: 2). Science appears in these analyses as a subculture with a diffused collection of institutions, areas of special knowledge, and theoretical interpretations whose forms and boundaries are open to negotiation with other social institutions and forms of knowledge (Irwin and Wynne 1996: 8). Such a redefinition of science has its roots in the sociology of scientific knowledge (SSK) tradition that emerged in the post–Thomas Kuhnian period of sociological/ethnographic studies on science.[13] Based on the insights from his classic case study of the Cumbrian hill sheep farmers' encounter with science, Wynne (1996)[14] argues that credibility of and trust in science are analytical tools themselves, which represent underlying tacit processes of social identity negotiations. Even the public 'ignorance' of science is a valid analytical artefact available to the researcher.[15] The contextualist model lays its emphasis on how publics *engage* with science in local contexts. While the deficit model blamed the public for lack of appreciation of science, the contextualist model established the agency of publics and demanded more reflexivity from experts to bring science in dialogue with them.

The contextualist model endeavoured to capture public engagement with science through micro-sociological case studies, employing participant observation and in-depth interviews as methods of data collection (see Wynne 1995). The publics the model attempted to capture, as mentioned earlier, were largely social groups and communities who had existed in specific contexts trying to articulate and defend their social identity as in the case of the Cumbrian sheep farmers (Wynne 1996: 39) or the villagers of the cyclone affected areas of Odisha (Dash 2014a, 2014b)[16] or the lesbian, gay, bisexual, and transgender (LGBT) activists of the AIDS (acquired immune deficiency syndrome) treatment movement in the United States who engaged with clinical trials (Epstein 1995, 1996) or the fishing community of St Brieuc Bay who collaborated with scientists to prevent a decline of scallop population (Callon 1986). Such a conceptualization of the public as a heterogeneous and socially dissipated network of citizen-collectives who engage with science in a variety of sites and contexts was a new sociological imagination of 'mini-publics' against the deficit model's attention to individual citizens, whose science consumption levels could easily be accessed through quantitative opinion surveys. In a later phase, the idea of the public became more linked to the theory of deliberative democracy that underscored the significance of public deliberation for deepening democracy.[17] Researching the role of public deliberation of science in risk regulation and technological governance has become a strong agenda for PEST researchers in the recent decades.

The deliberative turn, however, still continued largely with its sociological emphasis on concrete sites of construction of scientific citizenship and formation of publics. The studies on citizen science initiatives explored the possibilities for creating new institutional mechanisms and spaces for fostering citizen participation in technological governance such as citizen juries, consensus conferences, deliberative mapping exercises, science shops, and citizen groups engaged with scientific experts and policy makers ensuring better, democratically inclusive decision making regarding issues of technological governance and risk management.[18] It is worth noting that citizen science mechanisms have of late become an integral part of governmental science policy system in many Western countries.[19]

A major question that emerged at this juncture was about the very nature of engagement—especially the relationship between the mini-publics and the sites and mechanisms of participation. A second question was regarding the relationship these mini-publics have with the wider public, as suggested by Ulrich Felt and Maximilian Fochler

(2010). An overemphasis of the PEST researchers on such specific formal and informal micro-contexts of engagement seems to end up in under-theorization of the wider public and their linkages with mini-publics of both the 'invited' (formal) and 'uninvited' (informal, spontaneously generated) kinds.[20] In other words, although scholars have argued in favour of an intensified, unstructured, and informal public delibera-tion at large in society as quintessential to democratization of science and society,[21] the wider public it refers to have not been adequately subjected to nuanced theoretical treatment, other than Jasanoff's for-mulation of 'civic epistemology' in national contexts of engagement (Jasanoff 2005).[22]

Any theorization of a wider public cannot avoid the ubiquity of the mass media as a moulding force in our everyday practices and social relations (see Hepp 2012; Livingstone 2009). Political interventions now are impossible without being mass-mediated (Koopmans 2004). The phenomenon of 'mediatization' increasingly shapes politics in contemporary democracies and this has fostered the constitution of the wider public increasingly as a mediated public.[23] Mediatization of public engagement with science in contemporary risk societies is then a process that cannot be excluded from any attempt to theorize the wider, deliberative public.

Science and Media

The relationship between science and media has been a major focus of recent research in science communication studies. These investigations critique the earlier, linear model of communication (the transmission model), and proposes a multidirectional communication that exists between science and the public. According to Cloitre and Shinn (1985), there are four stages within science communication—intra-specialist, inter-specialist, pedagogical, and popular. At the intra-specialist level, scientific communication is made through specialist science journals and seminars and workshops in a particular area of research and at the inter-specialist level interdisciplinary communication is made through publications in 'bridge journals' like *Nature* and *Science*, and also through interdisciplinary seminars and conferences. At the pedagogical level, science textbooks and classroom teaching play the communica-tive role: scientific knowledge appears as epistemologically stable and coherent. At the popular level, scientific information is disseminated through mass media and popular science journals and programmes.[24] These different stages are not watertight compartments, but exist in continuum. Differences between the stages are only a matter of degree and their boundaries are porous if not open (Bucchi 1996, 1998).[25] What is important here is that each stage feeds back to the previous one unlike the proposition of the linear model. Therefore, it seems that all the four stages have their own functional autonomy and their own gatekeeping mechanisms, but are at the same time interconnected with other stages, guiding knowledge production.[26] As Bucchi (1996, 1998) points out, one stage does not necessarily lead to the next stage, further simplifying and stabilizing the knowledge available at the previous stage. Some research never reaches the pedagogical and popular stages. Also, scientists may directly approach the mass media with their research for several reasons ranging from immediate recognition and acknowledgement to winning over their rivals during scientific controversies. Mass media's peculiar role in scientific controversies further complicates this model of science communication—they substitute the conventional peer review process as a channel of intra-special communication between the rival groups of researchers (Lewenstein 1995a) and mediate 'the relationships among experts with the strong epistemological force of effecting what goes on in the laboratory' (Simon 2001: 383).[27]

The emphasis of these studies is on communication of science as a complex, multidirectional process that influences the cognitive process itself. This critical insight on the extra-transmissional function of communication channels has been further developed conceptually by a team of scholars who takes cue from the function systems theory of Niklas Luhmann.[28] Peter Weingart, the pioneer of this perspective, notes that there is a steadily increasing 'coupling' and mutual 'resonance' between science and media as part of a 'loss of distance' between science and different domains of its social environment (Weingart 1998, 2002). Science and media as two systems are structurally coupled today, a phenomenon known as 'medialization' which is part of a general orientation of science to other systems due to its internal differentiation.[29] Following Luhmann, Weingart argues that there is a 'close coupling between science and politics, the economy, the media, and law. These new arrangements have peculiar repercussions on knowledge production, on notions of true and false, certain and uncertain, and on the demarcations between science and non-science' (Weingart 2002: 170). The structural coupling between science and media is also due to a similar internal differentiation of media as a social (sub)system (Weingart 1998).[30] Media has developed its own criteria and parameters like profitability and public attention to represent the world and they construct their own reality in the same way as science does (Weingart 1998, 2002). Media's role has become more pertinent in shaping public opinion and public perception of reality; consequently, the monopoly of science in judging representational adequacy has deteriorated and science's abstract criterion of truth is now being confronted by the media's criterion of public acclaim (Weingart 1998, 2002).

Weingart argues that the importance gained by media in structuring public discourse has catalysed science's orientation towards media. Media provides space for science to communicate with a 'non-scientific public' and the recourse to the public serves the purpose of mobilizing legitimacy with reference to securing the expansion of the boundaries of science vis-à-vis its social environment, as well as settlement of conflicts within science (Weingart 1998). For Weingart the practice of pre-publication in the media (that is, instrumentalization of the media in order to attain priority and public attention) and the scientific community's attempt to achieve legitimacy through mass media in their competition for scarce resources are instances of medialization of science and of their close coupling. (Weingart 1998). Scientists are more inclined now to share their achievements with the media to increase their visibility in the public domain, hoping that this would help them secure more public funding for their research projects. Many of the top scientific journals also employ this strategy of announcing the research findings in a press conference prior to the publication of the research paper. Added public attention is expected here to attract more citation to the research article, eventually leading to a climb in the journal's impact factor. It has been noted that scientists who work on new fields of research such as climate science may initiate a 'catastrophe discourse' with the help of mass media to highlight the social significance of (and hence direct more funding towards) their area of research (Weingart 1998).[31] In cases of 'boundary work' as well, public support is indispensable to negotiate and constitute boundaries, and the engagement with the public through media is used as an effective means to attain this goal.[32] The novelty here is in the form and intensity that emanate from a closer linkage between science and its social environment as well as the new role of media in observing this connection (Weingart 1998).

The advent of new media accentuated the medialization phenomenon. The conventional peer review system of science is increasingly coming under pressure because of the posting of research findings on online platforms before being published as a journal article. This online sharing of research findings offers researchers better feedback from the peer group, and helps effective and prompt intra-specialist sharing of knowledge. New forms of intra- and inter-specialist communication emerge in the shape of email groups, blogs, and social media that boost the formation of virtual communities of researchers, with tremendous impact on the internal organization of the scientific community. Information and communication technologies (ICTs) in general have increased productivity by allowing immediate access to online resources and better possibilities of collaborative research with scientists from institutions across the globe. This also demands new skills and strategies from researchers for them to be successful.[33]

The accentuation of mutual resonance between science and media marks a changed configuration of science at the institutional and organizational levels.[34] However, scholars stress that this loss of distance in no way implies de-differentiation or blurring of boundaries between science and media (Franzen, Weingart, and Rödder 2012: 8; Weingart 2002: 181). Nevertheless, the gradual intensification of the interaction between science and media in recent times leads to tensions within science as this exposure to media is at odds with the rules and values of science (Weingart 2012: 24). This is further complicated by the erosion of the professional elites' long-standing privilege of being unaccountable to the public (Weingart 2012).

While diverse aspects of the system coupling between science and media are studied,[35] the theory curiously disengages from the concept of 'mediated publics' and reverts to a relaxed use of the term to denote a wider audience of science and/or media, or citizens in general. Research on medialization has its stress upon the changes in the internal organization of science rather than on the impact of the structural coupling on modes of public engagement with science. For instance, Franzen, Weingart, and Rödder (2012: 11–12) propose to explore medialization of science along three dimensions; these are: individual scientists' attitudes and adaptation strategies to media logic during interaction with the public, organizational responses of science, and its impact on scholarly communication (scientific publication systems).[36] These three levels of research on medialization show that their focus is on the internal reconfigurations of science rather than on the question of the publics, in the backdrop of the loss of distance from media that it has come to suffer from.[37]

Medialization theory perceives the public thus as a vague and general entity bereft of any specialized autonomy beyond their minimal existence as context-bound social groups or as imagined by the media.[38] For medialization theorists, 'the public, unless socially organised as in voluntary associations or semi-organised as in temporary demonstrations or town-hall meetings, only exists as an abstract "referent" of actions or communications' (Franzen, Weingart, and Rödder 2012: 8). Unlike science that is organized around concrete, organized communities of knowledge producers who constitute specific research fields and disciplines,

> [t]he mass media ... do not address communities delineated by criteria of inclusion and exclusion, let alone constituted by criteria of competence. Most commonly they address unspecific public, as large as possible because that usually translates into profits, the chief objective of the media as commercial enterprises. As public have become differentiated according to their interests, the media try to capture their attention by, first of all, researching their specific profiles: demographic, social structural, life style, etc. Although they may be very successful in 'profiling' their prospective audiences, *the public they address remains unknown to them in principle. (Mass) Media and their public are connected through contingent expectations rather than direct reciprocal communication.* (Weingart 2012: 20)[39]

From this vantage point, the scientific community is institutionalized with strict rules of exclusion and inclusion, while the media does not follow any such strict rules that constitute a community, making impossible the concept of a public that engages with science and media in their negotiation of risks. The public is virtually constructed by the media in reference to their audience and hence mass media can be the 'analytical proxy for the public' (Franzen, Weingart, and Rödder 2012: 8). The transformations of the wider public in a mediatized society is less attended by the medialization theory; the mediated public is theoretically less rewarding for them, being a virtual public constructed in the process of mediated communication of science to the audience. Mediated publics, hence, do not have a history; they are virtual, abstract, ephemeral, and contingent upon medialization alone.

Even though the risk society thesis points towards the importance of public sphere and mass media in risk deliberation, this has not been given enough methodological attention by Ulrich Beck (Cottle 1998). Nevertheless, the mode-2 thesis develops the concept of 'agora' as a structured public space populated with 'multiple publics and plural institutions' such as social movements and mass media that hosts public deliberations on science (Nowotny, Scott, and Gibbons 2001: 206). Agora still appears to be a dissipated space of engagement and its conceptual specifications are largely unaccounted for.[40] Besides, how it is conceptually different from and superior to the Habermasian concept of public sphere is mostly unclear, except for a cursory and indirect hint to agora's potential to go beyond the former as a social space 'in which market and politics meet and mingle, where the articulation of private emotions and meanings encounters the formation of public opinion and public consensus' (Nowotny, Scott, and Gibbons 2001: 183). On the other hand, the PEST framework develops the concept of multiple publics of civil society,[41] but flounders to register that civil society publics are increasingly operating in a media-saturated environment. Medialization theorists take into cognizance the nuances of science–media coupling, but they also lack a robust concept of mediated publics. It is also worth noting that the general audience of science and concrete 'ethno-epistemic assemblages' of publics (for example, the mass-mediated publics or the ones generated by social movements) are often conflated in these discussions.[42]

It is in this context that Rob Hagendijk (2004) offers theoretical clues to the shaping of a public at the intersection of science, mass media, and risk politics. He uses the concept of a 'regulated social world' wherein public controversies about S&T are manifested. The 'social world' defined as 'the construed and volatile order of everyday life and social experience' and 'regulation' is used 'to stress how everyday life around the globe has become subject to almost infinite, yet heterogeneous techno-scientific regulation, standardisation, and representation' (Hagendijk 2004: 54).[43] Everyday life has been under governance and coordination by means of different regimes of science including agriculture, food, health, education, industry, transport, and communication (Hagendijk 2004). According to Hagendijk, the diversity and heterogeneous nature of regulated worlds and their internal inconsistency and opaqueness demand 'agency' and 'responsibility' from individuals. However, much of the decision making related to these issues that affect everyday life eludes public review and hence the architecture and maintenance of our regulated worlds are monopolized by scientific, technical, and legal experts (Hagendijk 2004).

According to Hagendijk, mass media plays a considerable role in making these issues public, thereby constituting a mediated public of science by forming public opinion. He opines that most citizens start paying attention to such issues only when the media reports them; after that the systems of regulation become subject to public scrutiny and consequently involve public agencies along with scientific, technical, and legal experts. For him,

> [t]he mass media and related forms of communication provide the lenses through which the disempowered citizens of regulated worlds actually look at themselves. Regardless of the scepticism one may entertain with respect to the media as arenas for rational debate and informal communication, mass media are key vehicles of large-scale reflexive modernisation, turning citizens and consumers 'on' and 'off' according to new 'logics'. (Hagendijk 2004: 57–8)

If we extend this argument on the centrality of mass media in public engagement with science in risk society, it can be seen that the wider public itself exists as a mediated public, as the engagement is made possible by mass media. Hagendijk opines that three key factors have contributed to the emergence of this new mode of mass-mediated public–science engagement in liberal democracies—the changing configuration of contemporary science, the emergence of risk politics, and mass media's role in staging the public deliberations of technical decision making. Thus, Hagendijk clarifies that the mounting distrust and loss of authority of science is because of the changing institutional configuration of science and a subsequent mass media involvement in the scientific and technical issues that manifest in the regulated social worlds. However, what is the exact dynamics of the staging of this engagement that happens in mass media? How does the staging of public deliberations in mass media occur? Scholarly debates on public spheres in actually existing democracies offer a clue.

The Scientific Public Sphere

The idea of 'public sphere' can be a good starting point to conceptualize the emergence of a media-generated deliberative sphere (Habermas 1989). Jurgen Habermas identifies the emergence of 'rational-critical public debate', peculiar to the different national contexts of eighteenth-century Europe and the creation of a bourgeois public sphere existing through new social institutions like salons (in France), coffee houses (England), and table societies (Germany), wherein public opinion (*opinion publique*)[44] on political issues is formed through rational discourse (Habermas 1989). He marks the formation of a social practice of deliberation as the materialization of a bourgeois political public sphere.

> The bourgeois public sphere may be conceived above all as the sphere of private people come [*sic*] together as a public; they soon claimed the public sphere regulated from above against the public authorities themselves, to engage them in a debate over the general rules governing relations in the basically privatised but publicly relevant sphere of commodity exchange and social labour. The medium of this political confrontation was peculiar and without historical precedent: people's public use of their reason. (Habermas 1989: 27)

Thus the bourgeois public sphere was conceived as a sphere mediating between state and society, wherein public opinion played a role in the democratic control of state activity (Habermas 2000). Norms of reasoned discourse rather than social status or traditions decided the validity of arguments in this bourgeois political public sphere (Calhoun 1992: 2).

Habermas, nonetheless, is pessimistic about the potential existence of the public sphere in today's 'mass welfare-state democracies' when he describes disintegration, or the 'refeudalization', of the same, a process that began in the nineteenth century. The structural transformation of the public sphere led to the termination of the rational-critical discourse and it turned to be 'a field for competition among interests in the cruder form of forcible confrontation' (Habermas 2000: 293).

Thus, the public sphere was invaded by competition between organized private interests (Habermas 1989: 179). The public use of reason is shattered today as 'the public is split apart into minorities of specialists who put their reason to use non-publicly and the great mass of consumers whose receptiveness is public but uncritical' (Habermas 1989: 175). For Habermas then, the deterioration of the public sphere is triggered off by two crucial factors: the private use of reason in the public sphere against the Kantian ideal of the public use of private reason and the transformation of the critical public into consumers in the era of mass media and advanced capitalism. He points out several other reasons closely linked to the transformation of capitalism in the last three centuries as causative of the deterioration of the public sphere in contemporary societies, but our concern here regards to his emphasis on the emergence of mass media as a contributing factor. Although he acknowledges the origin of the bourgeois public sphere from its predecessor, the literary public sphere, and the role of the newspapers and magazines of the age that promoted rational-critical discourse, for Habermas the mass media in the advanced capitalist societies have to be blamed for the contemporary weakening of the public sphere. The early literary public sphere which was constituted by the press and an educated reading public collapsed with the advent of mass media that broadened and diversified the reading public to include almost all strata of the population as part of its commercialization (Habermas 1989). As a result of this process of expansion of the reading public, 'the press that submitted political issues to critical discussion in the long run lost its influence' and the political public was replaced by a culture-consuming public (Habermas 1989:169). Today's public sphere fashioned by the mass media, for Habermas, is therefore deceptive in its appearance as being actually transmogrified into a sphere of cultural consumption rather than a sphere of rational-critical discourse.[45]

The Habermasian pessimism towards the role of the mass media as a propaganda machine of powerful corporations and other interest groups of late capitalism is disputed by contemporary research on mass media. According to Kellner (2000), contemporary societies witness the expansion and redefinition of the public sphere as a site of information, discussion, contestation, political struggle, and organization constituted essentially through the mass media—an argument that goes against the grain of the elitist lamentation on the deterioration of politics in mass democracies.[46] The surfacing of new social movements in the 1960s and 1970s is a clear indication of the richness and vitality of contemporary public spheres (Calhoun 1992: 33; Kellner 2000).[47] Furthermore, as scholars have identified, in place of the bourgeois political public sphere defined by Habermas, there are multiple public spheres (and hence multiple publics) organized differently.[48] The Habermasian idealization of the eighteenth-century bourgeois political public sphere is thus countered by fresh theoretical reflections on the complexity and cultural peculiarities of the deliberative process in contemporary societies. It was also noted that the multiple public spheres may often be overlapping and mutually contending (Calhoun 1992: 37) or pushing new political agendas to the dominant public sphere (Fraser 1995, 2003).

The argument in favour of multiple public spheres in actually existing democracies raises the question of the possible existence of a public sphere organized around the rational-critical debate on science. Broman (1998) explores this possibility while enquiring how a public sphere came into being in the eighteenth century vis-à-vis the emergence of 'public criticism of science' and suggests that public debates over new medical and scientific theories in popular journals of this period were indicative of the development of the discourse.[49] Steven Shapin (1990: 1001) notes the 'face-to-face interaction in the seventeenth and eighteenth centuries between men of science and laypersons in public venues such as coffee-houses, taverns and exchanges' and the development of print as a vehicle of science communication to the laity. For Broman (1998: 144), the new discourse of public criticism of science 'fundamentally reconfigured the basis on which knowledge was considered authoritative' in eighteenth-century Europe. Although the relationship between science and the public in the eighteenth century was radically different from today's context of public engagement with science,[50] Broman's theorization of the existence of a public sphere that deliberated science opens up the possibility to conceptualize a 'scientific public sphere' in today's actually existing democracies.[51]

The concept of a mass-media-generated scientific public sphere seems useful as an analytical tool to study mass mediated practices of public engagement with science. Scientific public spheres of the actually existing democracies today have not become defunct; nor do they operate as originally conceived by Habermas. In actually existing democracies the (scientific) public sphere functions with active filtering mechanisms (Fraser 1995, 2003). Therefore, the concept is employed in this book not as an ideal one, but as a discursive mechanism that structures public deliberation on science and shapes the scientific-citizen publics. On the other hand, it helps us understand the historical processes that led to the emergence of public criticism of science.[52] This also offers a site to study the dynamics of democratization of science as well as an opportunity to assess the scope and constraints of liberal, deliberative democracy. It becomes highly significant at a time when the 'deliberative turn' in STS studies keeps the flag of Deweyian optimism flying high.[53]

While the structural coupling of science with media and politics is a wider rubric that refers to a range of structural transformations and internal differentiations, scientific public sphere becomes methodologically relevant for the present study more in the context of public controversies over science. A scientific public sphere can come into being during the eruption of a scientific controversy into the public domain, or when expert advice is sought during a public controversy. It should be noted that not all science reported in the media belong to the category of public controversy. Science reporting also includes regular popular science columns and television programmes that present scientific advancements to the public in the form of more stabilized and non-controversial information. In some contexts, media reports science as something spectacular. A good example is the reportage of a highly successful space mission, skilfully showcasing science for public consumption. There is also plenty of popular science in mass media that approaches science mostly uncritically, even though broader questions of ethical, theological, and philosophical hues are sometimes raised in connection with science. Public consumption of science through media in these forms generally does not trigger greater public participation and critical deliberation of science, but results in its 'passive' reception.[54] In spite of this there are instances of popular science articles acquiring new functional roles during public controversies, as the case studies in this book demonstrate.

Theatrical Structure of the Scientific Public Sphere

The scientific public sphere is assumed to have a theatrical structure as suggested by Fraser (1995). For her, it is an institutionalized arena of discursive relations, 'a theatre for debating and deliberating' (Fraser 1995: 287). The scientific public sphere has a front stage and a backstage like the modern theatre. Mediated public debate on S&T is treated as theatrical performance, taking cue from the dramaturgical tradition of Erving Goffman (1956), helping us examine how it is produced (at the backstage), performed (on the front stage), and critically appreciated (by the audience).[55] There are actors who appear and perform on the front stage and 'managers' at the backstage. The backstage management (partially) shapes the front-stage performance. And the theatre exists because of an active audience in front of whom the deliberation unfolds. It is the complex interaction between the audience, performers, and backstage managers that shapes the unravelling of the public controversies over science in the scientific public sphere. Thus, the model characterizes the deliberations in mass media as theatrical performances. To put it differently, it is the performative actions in the scientific public sphere that shapes diverse claims of truth, power, credibility, trust, identity, interests, and so on, during mediated controversies.

The theatrical structure of the scientific public sphere as with a front stage and a backstage consisting of different sets of players enables us to situate 'discursive practices in their socio-institutional context', unlike those approaches favouring 'a free-floating, decontextualized discourse analysis' (Goffman 1956: 288). Instead of constructing the deliberations on science by delimiting the analysis to selected news texts as generally done by media studies scholars, the attempt in the present work is to study the deliberation on the front stage as it unfolds in its richness on an everyday basis. The construction of the scientific public sphere in the image of theatre is helpful also in avoiding a theoretical predicament of the Habermasian conception of public sphere: his attempt to lay emphasis on the rational-critical discourse as constitutive of the public sphere in its strict form may not account for certain modes of communication specifically found in media-generated public spheres such as storytelling and rhetoric which are also decisive (Dahlberg 2005).[56] The conceptualization of the scientific public sphere as a theatrical space makes room for diverse renditions of the actors. This also helps us capture the performative strategies employed in the scientific public sphere.

In the theatrical structure of the scientific public sphere, the front stage and the backstage are two partially autonomous domains which symbiotically coexist and shape each other. The front stage of the scientific public sphere is maintained through the day-to-day reporting of the scientific controversy by the newspapers. Scientists, spokespersons of scientific and governmental bodies and organizations, individual citizens, popular science writers, social activists, politicians, public intellectuals, and so on, usually appear as actors of the unfolding controversy. The media organizations themselves participate in the debate through their editorials and special journalistic articles. Media audience may directly participate in the front-stage activities through strategies like letters to the editor. Journalistic production of news is an active process at the backstage that involves several players and interest groups. The stage managers such as journalists and scientists mobilize the resources to make the performance at the front stage possible, although they only have partial control over the complex unfolding of the drama on the front stage. The wider (mediated) public constitute an active audience who attend the theatre, being profoundly cognizant of the central role of the scientific public sphere in regulating their risk-filled social worlds.

***

Conceptualization of the scientific public sphere, as we have seen in this chapter, is methodologically helpful in capturing the social dynamics of public engagement with science in actually existing democracies. Public engagement with science, as the chapter showed, is deeply linked to science's relationship with other social systems, of which media and politics deserve special attention. Modern science gradually established a special contract with politics and the contract was founded upon a separation between truth and power. This separation led to privileging the technical over the political in the liberal democratic vision of Enlightenment as Yaron Ezrahi (1995) has suggested, and the public was understood largely as appreciative of the technical expertise that solves the problems of the social world. Scientific experts were given the authority to manage the problems encountered by the citizens and the nation-state. From the purview of the contract, media was expected to play the role of transmitting science to the citizens. Mass media connected science to their appreciative audience through popular science. However, this division started collapsing in the 1960s when an epistemological critique of science developed in the context of growing ecological awareness. This growing pessimism about science's role in politics, and the new advancements in science that created new crises for technological governance, led to the emergence of a public criticism of science that demanded the creation of new spaces and institutional mechanisms for public participation in the democratic governance of science. The rise of the risk discourse further accentuated public engagement with science.

These developments in the liberal national democracies of the late twentieth century created conditions for a loss of distance between science and other social systems. The most noteworthy among these has been the structural coupling of science with media and politics that has made the scientific public sphere a deliberative space to negotiate risks, a phenomenon that radically altered the characteristics of the scientific-citizen publics in the recent decades. The concept of a scientific public sphere and its theatrical structure, the chapter has argued, would methodologically help capture the dynamics of mediated public deliberations in contemporary risk societies and simultaneously develop a critique of the very limits of deliberative democracy.

Scientific public sphere as an idea is particularly employed in the book to investigate the transformations in public engagement with science in the South Asian context. An important question at this juncture is about the historical processes that led to the creation of a scientific public sphere in Kerala. What are the characteristics of the loss of distance science experiences with media and politics? What is the social history of public engagement with science in the region? What is the nature of the scientific-citizen public of Kerala that emerged through and further guided a specific form of engagement? These questions are pertinent especially because of the region's long history of public engagement with science. The public controversies examined in the book have transpired at a specific historical juncture that marked a shift in the civic epistemology of the region, taking effect from the emergence of the print media in the regional language as the prime site of public deliberation of science. The next chapter historicizes this transition that produced the scientific public sphere in Kerala in the 1990s.

Notes

1. For the transformations of Indian science in the twentieth century, see Raina and Jain (1997).

2. The relationship between modern science and liberal democracy is seen positively by some scholars like Robert K. Merton, while the same is being seen of late in a more critical light by political theorists like Yaron Ezrahi (1990). However, Ezrahi suggests a recent dissociation between science and liberal democracy due to social transition towards a postmodern condition. For a detailed analysis of various standpoints on the relationship between science and liberal democracy, see Wang (1999).

3. For a detailed discussion on the duality of truth and power, see Jasanoff (2003a), Wang (1999), and Weingart (1999).

4. 'Loss of distance' is an idiom used by Weingart (2002) to suggest this phenomenon that happens in contemporary society between science and many other function systems. On social systems and functional differentiation, see Luhmann (1995) and Stichweh (1992, 1996, 2015a, 2015b).

5. Emphasis removed from the original.

6. Emphasis as in the original.

7. Emphasis as in the original. See Shah (2011) for arguing the point in the context of genetically modified (GM) crops in India. The author argues that 'science-based risk assessment has inherent limitations, however rigorous, independent, and peer-reviewed it might be' (Shah 2011: 31).

8. For Ulrich Beck (2009: 95), subpolitics refers to 'the decoupling of politics from government' and the possibility of politics 'beyond the representative institutions of the nation-states' in risk societies (italics removed). To put it a bit differently, it is the possibility of the political to emerge from unconventional social sites outside the formal institutions of politics (Beck 1997: 53). Also see Holzer and Sørensen (2003) for a detailed discussion of the concept of subpolitics.

9. Emphasis as in the original.

10. See Gibbons et al. (1994) and Nowotny, Scott, and Gibbons (2001, 2003) for mode-2 science; and Ziman (1996, 2000) for post-academic science.

11. Consultancy science applies available knowledge to clearly defined technical problems. Normal science refers to the Kuhnian paradigmatic science.

12. It should be noted that the introduction to the English translation of *The Risk Society* was written by Brian Wynne, the main proponent of the contextualist model, along with Scott Lash.

13. See Kuhn (1996) for his ground breaking analysis of paradigm shifts within science.

14. Originally appeared as Wynne (1989). The case study explains how the scientists and the sheep farmers are different social groups attempting to express and defend their distinct social identities. The sheep farmers of the Lake District of northern England claimed that the radioactive fallout from the Chernobyl accident had contaminated their sheep flocks and upland pastures. The farmers

found their social and cultural identities threatened by the form of scientific interventions since the scientists were not willing to take into consideration the farmers' expertise on the local environment and their social practices and relationships. The farmers had expressed specialist knowledge for the conduct and development of science, but this was ignored by scientists in their experiments in the field. The scientists' ignorance, lack of interest in local realities, and imposition of false assumptions about the agency of local people ended up in a loss of trust among the farmers and the consequent failure of scientific experiments and predictions. The expertise of the farmers was based on a deep cultural outlook incompatible with the scientific–bureaucratic cultural idiom of standardization, formal and inflexible methods and procedures, and prediction and control (Wynne 1996).

15. Ignorance about science is 'not a deficit by default of knowledge', but actively constructed and maintained and it is part of the dynamic construction of social identity (Wynne 1995: 380). See also Michael (1996).

16. See Chapter 1.

17. See Dryzek (2000) for a detailed discussion on deliberative democracy.

18. A citizen jury usually consists of 12 to 20 random citizens who listen to the experts' arguments on the technical issue at stake through a series of sittings. The jury finally submits its recommendations to the government/ scientific community based on their assessment of the issue. In a consensus conference, experts explain various dimensions of the technological issue to a group of citizens. The citizens who participate in the exercise formulate critical questions on the basis of these inputs and the experts answer them. This exercise may continue for several rounds to enable a consensus. In deliberative mapping exercises, the panel of citizens discusses the technology options in several sessions in the context of a technical problem and finally interact with the experts in a joint workshop. Science shops intend to assist the citizen collectives in producing context-relevant research. For instance, a civic group can design a research experiment to test the level of pollution in drinking water with the help of the researchers at the science shop of the local university. For details, see Lövbrand, Pielke, Jr., and Beck (2011). For the concept of citizen science, see Irwin (1995, 2001). For a case study of public consultations about genetics in the UK, see Kerr, Cunningham-Burley, and Tutton (2007). For a study in the empirical context of Austria, see Felt and Fochler (2010).

19. For example, the House of Lords Report (UK) on *Science and Society* (2000). For a detailed analysis of this shift in science policy, see Irwin (2001) and Miller (2001).

20. See Felt and Fochler (2010) and Wynne (2007) for the concepts of mini-publics and invited and uninvited publics.

21. See Fischer (2005), Jasanoff (2003b), and Wynne (2003). See Durant (2011), and Lövbrand , Pielke, Jr., and Beck (2011) for a detailed exposition to the debate on channelizing the publics towards decision making in contemporary democracies.

22. Civic epistemology refers to culturally unique, historically shaped, and politically grounded, tacit and often institutionalized 'collective knowledge-ways' 'through which they [the public] assess the rationality and robustness of claims that seek to order their lives; demonstrations or arguments that fail to meet these tests may be dismissed as illegitimate or irrational' (Jasanoff 2005: 255).

23. For various aspects of mediatization, see Lundby (2009).

24. Rudolf Stichweh (2003) classifies the science communication system into interdisciplinary (that includes intra- and inter-special communication), pedagogical (same as the third stage in the continuum model), political (expert advice to government, research presentations before funding agencies, and so on), and general ('oriented towards the abstract publics of modern society') 'modes of popularisation'.

25. See also Hilgartner (1990) for a demonstration of the changes in the content of information passed by experts based on the nature of the audience they have.

26. The centrality of communication in knowledge production was well demonstrated by laboratory studies. Ethnographers of science have argued that writing a research paper is not merely reporting of the conclusions already arrived at from an experiment to the peers, but is integral to production of knowledge itself. See Knorr-Cetina (1981).

27. See also Neresini (2000) for a demonstration of the role of popular media in the construction of scientific facts.

28. A recent review of medialization theory also refers to the theory of social differentiation into 'fields' proposed by Pierre Bourdieu (1975). See Rödder (2011). Systems theory of Niklas Luhmann proposes that the functional differentiation in modern society has led to the emergence of different systems, which are autonomous and with specific 'operational codes' internal to them. These systems (such as law, media, science, politics, and religion) keep their distinctiveness and do not dominate each other, but may develop mutual dependencies. See Luhmann (1995, 2000), Stichweh (1996), Rödder, Franzen, and Weingart (2012), and Weingart (2012).

29. 'Medialization' is often used a little differently from 'mediatization'. More recent work suggests that while mediatization denotes wider processes of mediated communication and the social impact of media technologies in general, medialization refers to mass media as a social system (Franzen, Weingart, and Rödder 2012: 5), indicating that theoretically these conceptions

belong to two different traditions, although having a lot in common. Mike S. Schäfer (2014) points out that this is an ongoing debate within German media studies and suggests that '[b]oth camps differ somewhat in terms of their theoretical foundations and empirical paradigms, with the "medialization" scholars mostly relying on macro-theory such as general social systems theory, and the proponents of "mediatization" using interpretative, micro-sociological theory and methods' (Schäfer 2014: 572n2.), although this distinction is not always maintained in the research on science–media coupling. For the use of 'mediatization' and 'medialization' as nearly interchangeable terms, see Rödder and Schäfer (2010).

30. See Jeffrey (2000) for a similar analysis of the Indian press, although not explicitly based on systems theory.

31. The creation of environmental science as a new discipline is another example. See Weingart, Engels, and Pansegrau (2000).

32. 'Boundary work' is a concept that explains how boundaries are created and maintained between science and non-science and between scientific disciplines, specializations, and new theoretical traditions through negotiations. Boundaries can be restricted to exclude competing professional practices or, conversely, can be made flexible in order to incorporate such practices within the dominant framework (Bucchi 1996). For a detailed discussion on boundary work, see Galison and Stump (1996) and Gieryn (1995).

33. For a review of the impact of new media on science, see Schäfer (2014).

34. Scholars suggest that the degree of medialization varies among research fields and disciplines. See Schäfer (2009) and Rödder and Schäfer (2010).

35. See, for example, Weingart and Pansegrau (1999), Rödder and Schäfer (2010), Bocking (2010), and Rödder, Franzen, and Weingart (2012).

36. See also Weingart (1998: 878).

37. Two recent articles (authored by scholars belonging to the medialization tradition) that attempted to review a large band of studies on science and media interaction indirectly suggest this lack of emphasis on the mediated publics of science and the media generated public spheres which stage the deliberation. See Rödder (2011) and Schäfer (2014).

38. Although the public's existence in concrete situations such as social movements and public demonstrations are acknowledged, Weingart considers these as temporary and insignificant, without explaining why (see Franzen, Weingart, and Rödder 2012).

39. Italics added.

40. See Davenport and Leitch's (2005) discussion on the methodological aspects of agora which curiously lacks such specifications.

41. The concept of the public is theorized in the PEST in connection with the concept of civil society; that is, the context-specific mini-publics are a part of

the wider, abstract public constituted by the civil society. This imagined public is often understood as a national public, despite the civil society public being now conceptualized as heterogeneous, multiple publics. See Fraser (1995, 2003) for a detailed theorization of multiple publics. See Hess (2011) for the employment of the idea in the PEST literature. Elsewhere I have argued that the theorization of publics against the backdrop of civil society is methodologically inadequate to study those publics which do not belong to civil society such as the 'quasi-publics' and 'non-publics' (Varughese 2012, 2014, 2015b). See also the concluding chapter of the book.

42. There is great ambiguity in media theory in general regarding the ontology of audience and public. For the complexities of the problem, see Livingstone (2005). Ethno-epistemic assemblages refer to specific forms of alliances and networks which take shape in culturally specific contexts of public engagement with science (Irwin and Michael 2003).

43. His notion of 'social worlds' comes from the symbolic interactionist tradition in sociology.

44. For Habermas, public opinion can be formed only if a public that engages in rational discussion without coercion exists (Habermas 2000).

45. In Habermas's later writings also, the role of mass media in communication is not adequately discussed, perhaps due to his emphasis on the face-to-face interaction of everyday life as the model for his theory of communicative action (Calhoun 1992: 31).

46. See also McNair (2000) for a strong case of 'de-feudalization' of public sphere in contemporary mass democracies against the Habermasian pessimism regarding the 'crisis of public communication'.

47. Koopmans (2004) also underscores the linkage between contemporary social movements and the mass media. In modern democracies, the decisive part of the interaction between social movements and political authorities is indirect and mediated, and it takes place in the arena of mass media-generated public sphere. It has also been suggested that there is a neglect of social movements in Habermasian theory (Calhoun 1992: 36).

48. There are attempts to conceptualize specific public spheres in connection with various themes. For example, Johnson (1992) discusses the empirical existence of a musical public sphere in eighteenth-century France. Fraser (1995, 2003) conceptualizes the entire society as consisting of different public spheres. She calls the subaltern collectives as counter-public spheres, which act as alternative spaces for deliberating upon issues of their concern. She argues that the ideas thus created by the counter public in the respective counter-public spheres are finally pushed to the main public sphere and are further negotiated (Fraser 1995). She also discusses the possibility of a distinction between 'strong' and 'weak public' in which the former consists of decision-making bodies like

parliaments, but the latter consists of opinion-forming spaces like newspapers (Fraser 2003). Habermas himself hints towards the possible existence of many public spheres in *The Structural Transformation of the Public Sphere*, but focuses on the political public sphere (Habermas 1989). In a later discussion on his conception, he concedes his failure in conceptualizing alternative public spheres constituted by women, plebeians, and other marginal social groups in his original work (Habermas 1992).

49. See also Neidhardt (1993).

50. See Shapin (1974) for a detailed examination of the audience of science in the eighteenth century.

51. However, Broman contends that the scientific public sphere has become a hollow shell in contemporary societies à la the Habermasian pessimist view. For him, there has been a decline of public criticism of science in modern mass democracies—the public can neither establish nor dispute scientific truth in contemporary societies as a result of the growing internal differentiation of science into highly specialized disciplines. Scientists and physicians increasingly appear as specially privileged speakers of truth, and the public sphere has turned out to be a hollow shell today, although the shell looks intact and solid from outside (Broman 1998: 142). He argues that contemporary science artfully maintains the hollow shell of the public sphere to make the rhetorical claim that knowledge is open to public criticism, while maintaining its epistemological insularity by distancing from other social institutions and domains (Broman 1998: 142). My proposition of scientific public sphere in fact represents the breach of this epistemological insularity since the 1960s.

52. Nick Crossley (2004: 110) points out that a major drawback of the Habermasian theorization of the public sphere is the construction of an asocial model of communicative engagement that brackets out 'real world' conditions and contexts.

53. The American pragmatist philosopher John Dewey (1859–1952) was a pioneer in theorizing the importance of deliberation in democratic decision making, without which he felt that an elite class of experts would transform liberal democracy into a form of oligarchy. While his views on democracy remain highly relevant, it is also important to critically engage with them.

54. The term 'passive' is used here in a comparative context, in contrast with public participation during controversies. The reception of popular science is in no way passive as conceptualized by the deficit model. There are hardly any studies on popular science as an active site of engagement in the Indian context.

55. See Hilgartner (2000) for a study that employs the theatrical model to understand expert advice to the government and public in the United States.

See Bijker, Bal, and Hendriks (2009) for the employment of the model to study scientific advice in the Dutch context.

56. Dahlberg's attempt is to defend the Habermasian theory against the critique of the 'difference democrats' like Chantal Mouffe, Jane Flax, Nancy Fraser, and Iris M. Young. He argues that their critiques emanate from a limited reading of the Habermasian theory. See Dahlberg (2005).

3 Mediating Science
Scientific-Citizen Public and the Regional Press

Scientific public spheres are shaped differently in varied cultural contexts, and their formations are deeply linked to science's loss of distance with media and politics. In this chapter, we will be exploring the significant social and historical processes that made possible the emergence of the scientific public sphere in the recent decades in Kerala. The mediated space for public scrutiny of science was largely impacted by a shift of the site of public engagement with science from the Kerala Sasthra Sahithya Parishad (KSSP), the renowned people's science movement (PSM), to the regional newspapers in the late 1990s. This shift, as we shall see, is deeply linked to a transformation in the civic epistemology of public engagement with science (Jasanoff 2005).[1] The characteristic changes in the civic epistemology of the region in the new millennium, the chapter argues, were due to a radical shift in the relationship among science, media, and politics in the region. The recent emergence of their mutual resonance has a longer past in Kerala.

This chapter discusses the historical becoming of the scientific public sphere through an enquiry into the changing forms and discursive framing of public engagement with science in the region. The relationship between science and politics in Kerala is deeply linked to the politico-epistemological contract that came into being during the formation of the nation-state in the 1950s (Varughese 2012, 2015b). Emergence of a scientific-citizen public in Kerala was coterminous with the initiation of the contract, and this public was largely appreciative of the developmental project of the nation-state propelled by science and technology (S&T) (Varughese 2012, 2015b). The PSMs in the country originated in the backdrop of the contract, and the story of the KSSP reflects a transformation in the nature of public's engagement with science in the late 1970s. The emergence of environmentalism and the epistemological critique of modern science in the 1970s were crucial for this transition that defined the developmental modernity of the region. A second shift is visible in the recent surfacing of risks in the everyday social life of the citizens in the 1990s, wherein the regional media acquired a central role. Organization of the political culture around newspaper reading intensified since the 1960s, followed by a 'newspaper revolution' that was experienced all over the country in the 1970s (Jeffrey 2000, 2010). A unique kind of mediatization of politics took shape in the region in the decades that followed, which situated the newspaper-reading publics at the core of political culture as agents of political action. Medialization of science was but a more recent process, occurring in tandem with the acceleration of public perception of risk in the 1990s. These processes together have given shape to the scientific public sphere. The public controversies discussed in the subsequent chapters epitomize the scientific public sphere's maturation as the dominant form of public engagement with science in the new millennium.

Reader-Publics and the Political Culture

Kerala is known for its political culture which is deeply linked to the development of 'reading' as an everyday political practice. The region's reading habit has its roots in the nineteenth century, linked to the emergence of prose as a new mode of literary enunciation followed by the development of a print culture that existed along with reading as a new social practice. The advent of printing press and the simultaneous emergence of prose as a major literary form were important events which aided the emergence of the reading culture.

Printing press was introduced in Kerala by Christian missionaries in the nineteenth century and the early Malayalam journals like *Malayala Panchamgam* (1846), *Rajyasamacharam* (1847), *Paschimodayam* (1847), *Jnananikshepam* (1848), and *Vidyasamgraham* (1864) were published by them. The readers of these journals were mostly those who were close to the missions—a mixed group of Christian converts that mostly

ty contained the possibility of representing all kinds of interests, and

included the untouchable castes. The mission presses printed religious tracts as well as secular publications. A government press also was established by the princely state of Travancore as early as 1834 with the help of Protestant missionaries. The newly emergent class of intellectuals in the latter part of the century found prose as an efficient literary form to hold the modern ideas of the period, leading to the publication of novels and short fiction, textbooks,[2] and publications on general knowledge. Novels as a literary genre had a unique role in developing the dialogic mode of communication through imagined rational conversations about matters of public concern in the not-yet-arrived social spaces in the casteist society (Varughese 2015a). There were 33 magazines published between 1846 and 1900 in Malayalam (Sam 2003). The printing press started in Cochin in 1864 by Devji Bhimji (Keralamithram press), a Gujarati trader, and the Vidyabhivardhini Press in Kollam established in 1896 by S.T. Reddiar popularized books for casual (*nerampokku*) reading (Govi 1998). By the end of the century the number of Malayalam publications increased notably and newspapers like *Kerala Mithram* (1881), *Kerala Pathrika* (1884), *Kerala Sanchari* (1886), *Nasrani Deepika* (1887), *Malayala Manorama* (1890), and *Manorama* (1891)[3] commenced publication (Raghavan 1985). Arunima (2006: 69–70) opines that 'the shared world of readers created by the new journals [and newspapers] of the late nineteenth century contained the possibility of representing all kinds of interests, and created a space for debate and discussion in a manner unlike that of the [earlier] religious tracts'. She argues that by the close of the century, the emergent 'print culture' turned to be accommodative of a wide range of opinions and vantage points, conceiving a nascent public sphere.[4]

By the turn of the new century, newspapers in Malayalam became common, and literary production intensified, establishing prose as the dominant form of language, with the potential to articulate the modern, individual subject and his/her inner world (Ramakrishnan 2000: 485). The early twentieth century witnessed a steep increase and diversification of publications in Malayalam paralleled by a steady increase in literacy rate, associated with the spread of 'the desire for education' among the lower rungs of the society (Menon 1994: 144).[5] Between 1901 and 1925, 162 periodicals were published in Malayalam (Sam 2003). These periodicals addressed different reader-publics and discussed a variety of subjects. Sanal Mohan (2015: 124, 148fn) points out that in

Travancore by the 1920s Dalit organizations had started magazines like *Sadhujana Doothan* (1919)[6] and *Cherama Doothan* (1923)[7] to accelerate the Dalits' claiming of public space.[8] Similarly, there were several women's magazines which constituted a women reader-public, and the first among them had been published from as early as 1892 (*Keraleeya Sugunabodhini*) (Devika 2008: 50–1). Politics and social reform were the emphases for many of the periodicals and there were also several literary and non-literary publications which targeted the readers who sought personal enjoyment and religious satiation through the act of reading, as we have seen earlier.[9] By the 1930s newspapers became an important site for the constitution of a public sphere.[10] The newspapers as well as the magazines of the late nineteenth and early twentieth centuries had a dual role to perform—voicing the critical opinion of the public simultaneously also became an effort to produce a reader public(s) with the capability to engage in rational debate (Kumar 2007: 416).[11]

Many of the region's newspapers were started in the first quarter of the twentieth century to engage with the nationalist and reformist politics.[12] The spread of socialist and communist ideas and an eventual proliferation of the communist movement in Malabar and other parts coupled reading with mass politics.[13] Thus, by the 1940s, the political culture of the state was quintessentially popular in its orientation, thanks to mass participation in politics initiated by the peasant and workers' movements of the period (Menon 1994). The growing number of literates catalysed the establishment of reading rooms in every nook and corner of the region (Menon 1994). During the labour strike of 1937 at the New Durbar Beedi Company in Malabar, a major demand of the labour union was that '[w]hile the workers are idle without work, they should be permitted to read newspapers or books'.[14] Labour unions took initiative in the 1930s to establish reading rooms and night schools for workers (Isaac, Franke, and Raghavan 1998: 40). The activists of the Kerala Congress Socialist Party (KCSP) used the reading rooms and tea shops for propagation of leftist ideas, transforming them into spaces of political deliberation:

> One of the novelties in the organisation of the reading rooms was the communal drinking of tea, as one person read the newspapers and the others listened. Using literacy figures to determine the influence of a newspaper can be misleading, if only because newspapers were nearly always read communally. Tea and coffee lubricated discussions on the

veracity of the news and of political questions, and a new culture emerged around the reading rooms. It was premised on sobriety and knowledge rather than the drunken companionship transcending consciousness which characterised the toddy shops. The importance of tea and coffee lay in the fact that they were recently introduced beverages and did not fit into any taboos regarding what could be shared between castes. Tea shops and reading rooms all over Malabar provided a common place for people to meet and to drink together regardless of caste. (Isaac, Franke, and Raghavan 1998: 146–7)

Collective reading of newspapers thus became integral to the political culture in other parts of Kerala also by the 1940s. Robin Jeffrey observes that by the 1930s, tea shops were common in even the remotest villages of Cochin (Jeffrey 1992: 209–10). The practice of sipping tea, reading newspapers, and discussing politics was constitutive of a political public by the 1940s. The establishment of the reading habit was also catalysed by the library movement and the subsequent formation of the Grandhasala Sangham in 1945 (Pillai 2003). The movement catalysed the establishment of libraries in every nook and cranny of the state.[15] The social practice of reading proliferated in the following decades also with the advent of popular 'janapriya' literature.[16]

By the 1960s, the political culture of the region was subjected to a radical transformation with a new phenomenon that Robin Jeffrey (2000) calls the 'newspaper revolution'. According to him, the regional newspapers in India went through a 'revolution' in the 1970s in terms of mass circulation, adaptation of new printing technologies, and diversity of the news content as Indian capitalism entered a new stage of development. Although this was the national trend, the regional press in Kerala had started expanding in the 1960s itself. The rise in the circulation of newspapers was achieved through the conscious attempt of the newspapers to swell the readership, thereby increasing their revenue from advertisements (Jeffrey 2000).[17] This was accomplished by expanding the news coverage, reaching out to readers from all religions, castes, genders, and localities (Jeffrey 1997b). As a result, more local news content began appearing in the regional press. The press underwent some ideological depoliticization in their endeavour to boost circulation (Jeffrey 1997b). *Deshabhimani*, the newspaper owned by the CPI(M), best exemplified this trend. Jeffrey notes that the newspaper covered religious festivals and stock markets in the early 1990s, thereby departing from its earlier editorial policy more rigidly shaped by the Communist ideology (Jeffrey 2000: 134). The Total Literacy Campaign launched in 1988 by the state government[18] triumphantly accomplished its goal, and this further helped the expansion of the readership of newspapers (Jeffrey 2000: 19). By the year 2000, the total daily circulation of two Malayalam newspapers, *Malayala Manorama* and *Mathrubhumi*, was more than 2 million out of a total circulation of 2.36 million (Jeffrey 2000).[19] In 1957, the total daily circulation of Malayalam newspapers had been 246,000; by 1977, it had grown up to 1,169,000. By 1997 the figure doubled (2,427,000) and in 2001 their circulation almost touched 3 million copies per day (Jeffrey 2010: 273).[20] Since a single copy of a daily is read by five adult individuals on an average, the impact of newspapers has been much broader.

With this massive growth of Malayalam newspapers since the 1960s and the stabilization of this trend by the new millennium, even amidst the growth of television channels and, of late, of social media with the advent of internet and smart phones,[21] newspaper reading has become an essential part of the everyday life of Keralites.[22] Jeffrey (2000: 19) suggests that the everyday practice of newspaper reading deepened the politicization of society:

> As the newspaper [reading] habit spreads, the prevalence of newspapers, and their keenness for local news, opens up new avenues of politics. In short, the desire to know, created by political involvement, provokes a strong demand for existing newspapers. In response to that demand, proprietors expand their newspapers, using the most effective technology available.... Newspaper proprietors used advertisement revenues, swelled because larger circulations promised larger markets, to pay for the technology that enabled still further expansion. This expansion in turn built a vastly enlarged public arena, open to many more people than ever before. The nature of political processes was changed thereby.

The creation of new avenues for politics by the newspaper revolution catalysed the expansion of the political public sphere in the last quarter of the twentieth century. The widespread culture of newspaper reading that enabled the common people in Kerala to participate in everyday politics shaped the political public sphere as primarily constituted through the Malayalam press.[23]

However, as Jeffrey points out, the press does not produce the idealized Habermasian kind of public sphere of rationally debating citizens, but 'it does produce political conditions fundamentally different from those existing in relatively unconnected towns and villages among non-reading populations. Newspapers lay out avenues for public activity that previously did not exist' (Jeffrey 2000: 18). The expansion of the regional press thus created a public sphere and a public that actively engaged in political deliberations. In opposition to the Habermasian apprehension regarding the mass media generated public spheres in contemporary mass democracies, Jeffrey presents it as productive, wherein the public is not easily manipulated by propaganda. According to him, 'The messages of [newspaper] proprietors and their class are capable of being accepted, rejected or reinterpreted by readers. Readers, moreover, are capable of injecting their messages into print and push-ing their concerns to the fore, rather than simply being pushed in any direction proprietors care to guide them' (Jeffrey 2000: 13).[24]

The massive proliferation of the habit of newspaper reading since the 1970s and its stabilization as an everyday social practice suggests that even today newspapers in the region predominantly constitute the pub-lic sphere. However, there is a radical difference in this cultural practice from the early decades; now newspaper reading is a quintessential part of individual life. Collective reading practices which had shaped the political culture are gradually dying out with a steady disappearance of public libraries, reading rooms, street corner meetings, youth clubs, and political demonstrations in recent times, leading to a changed notion of the public.[25] However, this has increased the political importance of newspaper reading in a different but unique way—now the practice has greater significance in constituting the political public, as the individu-als increasingly participate in politics *primarily* by reading newspapers in an era of mediatization of politics.[26] The political public of Kerala hence is a newspaper-reading public.

Shaping of the Scientific-Citizen Public

Formation of a scientific-citizen public in Kerala is deeply connected with this social history of reading as constitutive of the political cul-ture. A new form of public engagement with science emerged in the 1970s in the region and the KSSP was the most prominent medium of its actualization. Earlier forms of public discourse of science in Kerala seem to have been veered towards domestication of science through its cultural redefinition (cf. Raina and Habib 2004). The early magazines in Malayalam showed great enthusiasm in presenting new developments in modern science and technology to their readers (Vadavathoor 2001). *Paschimodayam*,[27] the second news magazine in Malayalam, was a popular science magazine that introduced the 'rise of the West' ('paschimodayam') with the help of science and technology to its small circle of readers in Malabar.[28] The textbook committee of Travancore (1867) also was crucial in publishing many science textbooks for school children. By the early twentieth century, publishing popular science became an important objective for most of the magazines and newspapers, cutting across their specializations and target audience (see Vadavathoor 2001). The presence of a large number of popular science magazines and books since the mid-nineteenth century suggests that the public was keen on engaging with science during the period, although a dearth of research on the subject prevents us from understanding the nature of this engagement.[29] Science popularization underwent a major transformation in the 1960s, fostering a critical engagement with science beyond scientism. The KSSP and similar organizations in the 1970s and 1980s provided impetus to this cultural trend by aiming at the democratization of scientific decision making beyond mere popularization of science.[30]

The KSSP's origin can be linked to at least three separate initiatives to kick-start systematic publication of popular science in Malayalam. In 1957, a group of left-leaning science writers started an organization to facilitate popular science writing called the Sasthra Sahithya Samithi (Science Literary Forum) at Ottappalam, in Palakkad district.[31] However, they could not sustain it for long. In 1962, a similar forum was organized in Kozhikode named the KSSP, and some of the members of the former venture became part of the new initiative.[32] The members of this fresh initiative were mostly scientists who felt it was their responsibility to communicate science to the public at large.[33] Dr K.G. Adiyodi was the leader of the organization that initially had 30 science writers as members. Apart from its foremost objective of publishing popular science in Malayalam, the organization also planned to conduct public lectures, seminars, and film shows to cultivate enthusiasm and awareness of science among people (Zachariah and Sooryamoorthy 1994: 54). A third initiative that later became linked with the former came from a group of USSR-trained scientists and engineers in Bombay (now known as Mumbai) in 1965, who found it part of their vocation to impart scientific knowledge for the benefit of the Indian population.[34] They started a series of organizations to popularize science in their mother tongues, and the Sasthra Sahithya Parishad formed under the leadership of Dr M.P. Parameswaran in 1966 targeted the Malayalam-speaking audience.[35] The new organization soon developed connections with the KSSP.

The KSSP was formally inaugurated at Kozhikode on 10 September 1962. The main inspiration behind the organization was a common interest in spreading scientific values and perspectives as expressed by Dr K.G. Adiyodi in the inaugural meeting (also note his strong conviction in the scientific community's exclusive authority over science):

> We are living in an age of science.... The general laws of science and the discoveries of science, that exert such great influence on the development of mankind, should remain as the family property of certain experts. Either these experts themselves or some other people have to shoulder the responsibility of explaining them to the common man in a language that he can understand. Because of the step motherly attitude to the 'native languages' during the period of British rule, there was hardly any progress of science literature in various Indian languages.... On the other hand, the conservative local 'sastries' insisted that we have a science created by the great sages of the past that cannot contain any mistakes. Consequently, the awareness of the people of our country lagged behind that of others. Only regional languages can enter into a dialogue with the heart of the common man. The task before the science writers in Malayalam as well as in other languages is to convey the message of the new knowledge to the hearts of the people in a style they can easily understand.[36]

While communicating science to the public through science-popularization initiatives in Malayalam remained the shared interest, there was not much internal consensus on the nature of social relations of science. At least some of its members had strong socialist/Marxist inclinations and argued for the organization's involvement in society beyond the limited agenda of science popularization, from the vantage point of 'social function of science', proposed by John Desmond Bernal, British scientist and Marxist historian of science.[37] Members like Dr Adiyodi, as we have seen previously, argued for a strict separation of science from politics and its elevation as a special activity under the custodianship of experts. They wanted the organization to solely undertake science popularization. Although the radicals (Bernalists) were a minority in the initial years, gradually they became stronger and by the third annual meeting at Olavakkode (1966), the social function of science beyond mere science popularization became a serious point for discussion. The interventions of the Bombay unit must have facilitated the development of the Bernalist perspective. The internal debate continued for almost a decade and it was only in December 1973 at the 11th annual conference at Thiruvananthapuram that 'science for social revolution' was declared as the KSSP's official ideology.

During its first decade (1962–73), the KSSP had a moderate agenda of popularization of science, and this was executed through popular science publications,[38] organizing public lectures, seminars, and campaigns on science, initiating a discussion on creating a scientific terminology in Malayalam, and starting science clubs in schools (see Isaac and Ekbal 1988). The activities of the organization at this juncture reflected the (Nehruvian) nationalist spirit of the times that linked science with nation building. Scientists, engineers, and the educated elite had to play a nurturing role as mediators between the nation-state and the population groups to ensure the progress of the nation in the path of modernization and the activities of the KSSP harboured this spirit. By 1969, the organization had 500 members consisting of scientists, science writers, and school and college teachers who were engaged in popularizing science (Zachariah and Sooryamoorthy 1994).

The annual meeting of 1973 that accepted science for social revolution as the official perspective of the KSSP attempted to reconsider its understanding of the public as ignorant-masses-to-be-scientifically-tempered. The public was now redefined from a Marxist point of view as belonging to the lower economic classes—the rural population and labourers. The KSSP realized that everyday activities of the Indian population who belonged to the working classes had a scientific content, which had to be acknowledged and linked with the formal activity of modern science (Parameswaran 2008: 45). Following this transition in its understanding of the public as knowledgeable and productive from an earlier notion of them as common masses with a deficit of scientific knowledge, the KSSP started its village science forums (Grama Sasthra Samithis).[39] At the same time, the organization continued to support the Nehruvian developmentalist vision for the nation. The first activists' camp of the KSSP at Peechi in 1975 further consolidated its standpoint on harnessing the potential of S&T for development. And the camp found it important to take up the development of Kerala as a major concern for the organization.[40] Following the debate, Dr K.G. Adiyodi, who had been strongly representing the position that science must continue to be a special activity vested upon the scientific community and safeguarded from politics, resigned from the organization (Parameswaran 2008: 30). This was the same year (1975) that Dr M.P. Parameswaran, who later became the most recognizable face of the KSSP, left his nuclear engineering job and joined the organization as a full-time volunteer. The resignation of Adiyodi and the entry of Parameswaran (who was a strong votary of the Marxist–Bernalist standpoint) thus turned to be highly metaphorical of the organization's ideological transition.

Since 1976, the KSSP has been actively participating in the developmental discourse that was taking shape in Kerala. The KSSP by then had become the harbinger of the Nehruvian developmentalism in the region. The 1970s were marked by the creation of a large number of scientific and technological institutions as well as technoscientific projects in Kerala, actualizing the developmental aspirations of the region.[41] The KSSP had raised three main concerns at this juncture: the pressing need for industrialization for the region's development; the significance of mapping Kerala's natural resources to assess the region's industrialization potential; and, third, the organization endeavoured to harness the knowledge and skills potential of the productive classes for enhancing development. The programmes and activities of the KSSP were organized around these three concerns that emerged from its Nehruvian–Marxist–Bernalist viewpoint on the interface between science and society.[42] Unlike the earlier perspective on science as the privilege of the scientific elite, by the mid-1970s, the KSSP reclaimed it as belonging to the wider population. A critical discourse on science became possible with the adoption of the Marxist perspective: science was no more separated from politics and its central role in serving capitalist forces became explicit through the Bernalist lenses. Criticism of scientism as an ideology that distanced science from the public became part of the developmental discourse of the period. At the same time, the cognitive core of science continued to be understood as insulated from the social. Science was upheld and appreciated as the highest truth-seeking method but the need for decoupling it from capitalism to be used for social transformation was equally asserted.

Many of the concerns of the previous phase continued in the new phase, such as its emphasis on science popularization, but it was now neatly fit into a wider framework of harnessing S&T's epistemological potential for the socio-economic development of Kerala. The KSSP actively facilitated science popularization through its publications. The strategy of targeting school children as well as its art caravans (*sasthra kala jatha*), which received wide public attention, immensely helped in propagating popular science literature in the 1970s and 1980s. As the political culture of the region had been organized around the everyday practice of reading, the KSSP's science popularization efforts struck a chord with the public, and popular science became a popular genre for reading.[43] The KSSP's endeavours to link S&T with the developmental debate helped in placing the discourse on social relations of science at the core of political debate.

The Nehruvian–Marxist–Bernalist ideological orientation of the KSSP must be situated in the context of the politico-epistemological contract between the postcolonial Indian state and science to get a more vivid picture of the new mode of public engagement with science that emerged in the 1970s and 1980s. The Nehruvian developmentalist state safeguarded science from political interventions as an arcane institution enjoying high epistemological and moral status for being the engine of progress. In return, science legitimized the former's developmentalist rhetoric (Nandy 1988; Raina 1997). This symbiosis between science and the state was central to the political imagination of the citizenry being appreciative of the developmentalist trajectory of the Indian state with science as its epistemological vanguard. This learned and enchanted scientific-citizen public was constructed in contrast to the majority of the population who should be equipped with a scientific temper through popularization programmes, so that they could also appreciate the developmentalist project of the nation-state that promised a brighter and progressive future for the nation.

However, by the late 1970s, both the elite scientific-citizen publics and the population groups in India increasingly became sceptical of the developmentalist project of the nation-state and the political and epistemological function of modern science as its epistemological engine (see Raina 1997). This was also a time when the social contract between science and state was challenged by grassroots groups and dissenting academics who articulated a critic of science.[44] The emergence of a wide range of new social movements in the period indicated a growing apathy and disillusionment regarding the technocratic vision of development of the Nehruvian era. The crisis was aggravated in the context of the imposition of the state of Emergency (1975–7) by the then prime minister Indira Gandhi. Visvanathan and Parmar (2002) suggest that the Emergency was a period of dictatorship legitimized symbolically by science. 'Not only was the violence and demolitions and sterilisation, justified in the name of science, the emergency was seen as a scientific project. It brought in its aftermath a virtual explosion of debates on science and technology' (Visvanathan and Parmar 2002: 2714). The first nuclear experiment in Pokhran (1974), the imposition of forced family planning, the slum eviction drives, the forest policy that displaced the forest-dwelling communities, and so on, enhanced the collapse of scientific optimism of the Indian citizens. The demonic face of modern science and the associated development paradigm were further revealed in the following decades by the new social movements that emerged following the Bhopal Gas Tragedy (1984) and mass displacement caused by developmentalist projects. Dams, nuclear reactors, uranium mines, and forest management projects were opposed for their adverse impact on local environments and people from diverse vantage points.[45]

It is against the backdrop of these wider changes in the public discourse on science in the 1970s and 1980s that the relevance of the KSSP must be examined.[46] The origin of the KSSP can be traced back to the endeavours of different groups of elite scientific-citizen publics who took the mantle of propagating science and the liberal Enlightenment values and vision it had embodied to the wider population. However, as we have seen, from the mid-1970s onwards the KSSP developed a critique of science's social relations in a capitalist society, and proposed people's reclaiming of science for social transformation. This ideological shift in the organization harboured a new kind of scientific-citizen public that engaged with science more critically, but still largely convinced about its epistemological supremacy in providing technical solutions to the problems of society.

A strict separation of truth and power was but at the core of the KSSP's supple standpoint on 're-politicization of science'. In the late 1970s, there were several ecological groups and intellectual collectives in Kerala offering a more radical epistemological critique of science. They pointed out the role of modern (Western) science in pushing the world down the road of an ecological crisis. In the 1980s, these perspectives were strongly articulated in the public sphere, although the KSSP snubbed these views altogether as a 'naive' and 'romantic' environmentalism influenced by M.K. Gandhi and E.F. Schumacher.[47] The KSSP, being loyal to the Nehruvian–Marxist–Bernalist perspective on the social relation of science, considered these epistemological critiques were 'irrational' and 'unscientific'.[48] The Silent Valley struggle,[49] the first major public controversy that epitomized the changes in the discourses on science and developmental modernity, forced the KSSP to frame its 'scientific' perspective on environment as opposed to the views of the ecological groups.[50]

The KSSP's position against the hydroelectric power project in Silent Valley was articulated only in 1979, following a long debate within the organization.[51] The report of the commission appointed by the KSSP in 1977 to study the proposed dam project was published in 1979, paving the way to the development of its official standpoint.[52] The debate during this period clearly demonstrates that the KSSP's perspective on environment was based on the principle of cost–benefit analysis.[53] The KSSP decided to oppose the dam based on expert analysis of the Silent Valley hydroelectric project conducted by various scientific institutions. The technological project had to be opposed because of environmental, technical, economic, and socio-political reasons, and the KSSP upheld its 'integrated approach' to environment based on rigorous scientific assessment against both the scientism that informed developmental decision making so far, and the deep ecological perspectives of environmental groups. It pointed out that the ecosystem of the tropical rain forest of Silent Valley was unique and extremely rich in its genetic diversity (*gene sampath*), which was going to be a 'standing-reserve' for humankind's use in the future (cf. Heidegger 1977). The project would not solve the energy crisis in Malabar due to poor developmental planning, it pointed out. Kerala can be energy sufficient only if more efficient thermal power plants were installed instead of large hydroelectric power projects, an argument that the KSSP would develop into a campaign for the thermal power plant in southern Kerala (Kayamkulam) in the 1990s.[54] The KSSP also demanded immediate completion of the other hydroelectric projects under construction to reduce the power crisis.[55] These projects, in the KSSP's opinion, had lesser ecological impact and better benefits than the former. To resolve the power crisis of Kerala, it proposed efficient, long-term planning (*asoothranam*) that was based on scientific assessment.

Its standpoint on the Silent Valley project suggests that the KSSP's ecological view was founded upon a strong belief in science's epistemological prowess in providing solutions to societal problems. Despite being a late entrant to the Silent Valley movement, the KSSP soon became its public face. With its involvement in the struggle, the KSSP grew into a mass movement with a steep increase in membership from 5,859 in 1981–2 to 37,653 in 1988–9 and the number of units escalated during the decade from 300 to 1,195 (Zachariah and Sooryamoorthy 1994: 139). The majority of its membership came from Left-inclined educated middle classes/castes, mostly male (Zachariah and Sooryamoorthy 1994: 138–9).[56] By the early 1980s the KSSP transfigured itself into a successful social movement that radically shaped the civic epistemology of Kerala. It became the main conduit for public engagement with science in the region under the agency of a scientific-citizen public who enthusiastically defended the Nehruvian–Marxist–Bernalist perspective of a 'people-oriented science' against both the positivist philosophy of science that had informed developmental planning, and the epistemological critique of science proposed by new social movements.[57]

Risk Discourse and the Medialization of Science

What happened in the 1990s was a radical break from the mode of public engagement established through the KSSP in the 1980s. This shift in civic epistemology was in the context of manifestation of modernization risks in the everyday life of the public. These risks were unforeseen and they defied scientific analysis. In sharp contrast to the developmental issues of the previous decades, risks affected the everyday life of the citizens, demanding immediate intervention from government and experts. This has led to a serious epistemological crisis—there was no conclusive expert advice available to the public. The technical advice was often ambivalent and contradicting with the risk perceptions of the citizens. The conventional methods of scientific regulation of technical issues became futile in the new context. This crisis broke open a fertile ground for the spontaneous development of public controversies over science, hosted by the regional press: a scientific public sphere thus came into existence through the Malayalam newspapers.

The newly developed risk controversies were radically different from the earlier controversies like the Silent Valley dispute that were primarily mediated by the KSSP and other civil society organizations. The media only had a secondary role in reporting the events and debates to the reader-public in the previous phase. The KSSP had its own publications to communicate the issues to the wider public and more direct communication with citizens was accomplished through public lectures and art caravans. The public controversy over the hydroelectric dam project in the Silent Valley forest had materialized around the technical decision to build the dam to solve the energy crisis, and the scientific-citizen public argued that the state failed to pay heed to expert advice. The expert advice upheld by the state was found untenable and redundant due to its lack of methodological attentiveness to environmental and socio-economic costs of the project. Following the public debate, a consensus was reached and the project was abandoned due to lack of technical feasibility. The new disputes were not of a similar kind; they were risk controversies which demanded expert advice of a much wider range to enable the citizens to regulate their social worlds, while a consensus was not easily plausible due to the lack of any predetermined scientific method for technical assessment. While the KSSP was the arbitrator of the developmental disputes of the previous period, a wider, open platform was now required for risk deliberation, and the regional press readily extended such a space that could accommodate a wide spectrum of actors and perceptions. The civic epistemology of the region thus transitioned from an earlier *consensus-seeking* mode to a new, *contentious* mode in the 1990s onwards with the emergence of the scientific public sphere, as the public controversies analysed in the following chapters demonstrate.[58]

One of the earliest examples of a public controversy of science in the new phase was an unidentified microbial infection that killed fish in the water bodies of the region during 1991–2. Suddenly schools of fish were found dead and floating in paddy fields, backwaters and rivers, and this created great public anxiety because of the cultural status of fish as an essential part of the dietary habits of Keralites across classes and castes. Experts failed to provide a convincing scientific explanation for this phenomenon and the regional press hosted and actively shaped the public controversy that followed. Similar controversies over different kinds of risks kept appearing in the 1990s exposing the complex and dense tangle of truth and power, and science and politics to the public.[59] Such controversies have gradually become quite frequent and eventually made the old mode of public engagement with science represented by the KSSP redundant. In the new millennium risk controversies have become an everyday affair, and often several controversies ran parallel to each other.[60] The entry of mass media as the prime site of PEST in Kerala in the 1990s produced a new scientific-citizen public and altered the civic epistemology of the region. The present study focuses on this transition in civic epistemology in Kerala in the 1990s in tandem with the emergence of a mediated scientific-citizen public.

***

In the present chapter, my attempt was to examine the historical processes that led to the emergence of a specific form of public engagement with science in Kerala in the 1990s, which was fundamentally different from the previous period. The concept of the scientific public sphere provides us with a methodological frame to understand the public deliberations over science mediated predominantly by the regional press. The emergence of the scientific public sphere in the 1990s was made possible through the mutual resonance of science, media, and politics and the chapter enquired the historical processes behind the loss of distance among these social (sub)systems. Politics started becoming mass media oriented in Kerala in the 1970s following the emergence of a newspaper-reading publics at the core of the political culture. By the 1990s this process had aggravated. The KSSP during this period negotiated the possibilities for democratization of society with the help of science. A structural coupling between science and politics that emerged in the 1990s through the subpolitics of risks collapsed the neat division of the technical and political, nature and culture, and, more fundamentally, truth and power. The manifestation of risks in the everyday life of the citizens also facilitated a loss of distance between science and media, for the scientific-citizen public demanded the scientific community to be more engaging with mass media which seemingly reshaped science.

The scientific public sphere, as we have seen in the chapter, emerged as an unconventional site of politics (cf. Beck 1997) at the confluence of these historical processes specific to the region.

The scientific public sphere that is produced through the mutual resonance of science, mass media, and politics in Kerala is a site to look for the complex dynamics of public engagement with science that shapes the civic epistemology of the region in contemporary times. The three controversies discussed in the following chapters will help us capture the characteristics of the scientific public sphere and the way the debates have evolved. Analysing the deliberations will also help us understand the shaping of the scientific-citizen public in the post-1990s and the central role of mass media in the process. The case studies will give insights into how trust in scientists and credibility of scientific institutions were renegotiated during risk controversies. In short, the public deliberations on science staged by regional media opens a window into how public contestations of knowledge shape science, media, and politics in contemporary liberal democracies.

Notes

1. 'By turning to the concept of civic epistemology', Sheila Jasanoff opines,

   we move away from a priori assumptions about what publics should know or understand of science. Instead, we pose the question, salient in any democracy, how knowledge comes to be perceived as reliable in political settings, and how scientific claims, more specifically, pattern as authoritative. Put differently, civic epistemology conceptualises the credibility of science in contemporary political life as a phenomenon to be explained, not to be taken for granted. In shifting attention away from individual knowledge or ignorance of facts to how political communities know things in common, the concept of civic epistemology also offers a means of getting at cross-cultural diversity in public responses to science and technology. (Jasanoff 2005: 250)

2. The first set of textbooks was published by the Travancore textbook committee (1867).

3. *Manorama* did not have any connection with the *Malayala Manorama* published from Kottayam. The former was a newspaper published from Kozhikode and hence popularly known as *Kozhikodan Manorama*. See Raghavan (1985: 81).

4. However, since its inception, the nascent public sphere was exclusionary in nature, making possible only certain kinds of public enunciations. See Varughese (2015a).

5. One thousand and four elementary schools for boys and 317 for girls were functional in Malabar alone, providing education to approximately 18,000 pupils by the early 1930s. See Menon (1994: 144).

6. Started by Ayyankali from Thrikkodithanam, near Changanacherry.

7. Published fortnightly under the leadership of Pampadi John Joseph.

8. It is also important to note that Dalit leaders like Poikayil Yohannan of the Prathyaksha Raksha Daiva Sabha (PRDS) asserted the need for written history of the community, realizing the power of written word in constituting Dalit subjectivity (Mohan 2015).

9. See also Kumar (2013).

10. The idea of the public was produced also through diverse deliberative sites like reading clubs, public demonstrations (jatha), public meetings, and voluntary organizations in the early twentieth century.

11. Feminist scholarship on the period points out that the public sphere thus formed was masculine in its membership and deliberative functions. See Devika (2007) and Devika and Sukumar (2006).

12. The social reform movements of the turn of the century had profoundly determined the emergent political public sphere in Kerala. The reform movements included the Sree Narayana Dharmaparipalana (SNDP) Yogam (1903), Sadhujana Paripalana Sangham (1905) established by the Dalit leader Ayyankali, and the Namboodiri Yoga Kshema Sabha (1908). The Vaikom Satyagaraha (1924–5) and the Guruvayoor Satyagraha (1931–2) were also of great political significance for the nationalist politics in the region.

13. A note to the non-Indian readers of the book: The Communist Party was in power in Kerala from 1957 to 1959. The ministry led by E.M.S. Namboodiripad as the chief minister (CM) is often hailed as 'the first elected communist government anywhere in the world'.

14. *Mathrubhumi*, 30 November 1937, quoted in Isaac, Franke, and Raghavan (1998: 40).

15. Kerala has the largest number of rural libraries per head in India (Pillai 2003: 105). There are 4,000 public libraries attached to the Grandhasala Sangham and the total number of libraries in the state is many times larger, as most of the small libraries are run by voluntary associations (Pillai 2003). According to an estimate, there are approximately four libraries per village (Pillai 2003).

16. While the little magazine tradition corresponded with the creation of an elite reader-public, the weekly magazines published from Kottayam inaugurated a more popular and vibrant readership in Kerala. The major magazines which published pulp fiction (derogatively called 'painkili' sahithyam) were *Manorama Weekly* and *Mangalam*; together these two popular magazines sold 2.3 million copies a week in 1993, according to Jeffrey (2000: 101). For the

nuances of popular literature and reading habits in Malayalam, see Bay (2015) and S. Jacob (2009).

17. See Jeffrey (1997a) for a detailed discussion on the role of advertising in the evolution of Indian-language newspapers.

18. The literacy movement began in 1988 by the launching of a Total Literacy Campaign at the state level after the grand victory of the Literacy Programme in Ernakulam district. Kerala was declared a completely literate state on 18 April 1991. The KSSP was a major partner in the programme.

19. These two newspapers together hold 87 per cent of the total readership in Kerala, according to the 2002 January–June figures of Audit Bureau of Circulation (ABC). See the author's introduction (to the second edition, 2003) in Jeffrey (2000).

20. In 2011, the daily circulation reached 3.74 million. The share of *Malayala Manorama* and *Mathrubhumi* in daily circulation has increased to 88.75 per cent with a rising number of editions. Currently both of them have 20 editions each. These two newspapers continue to be in the top-10 list of the most circulated dailies in India. See the *Review of the Press in India (2008 to 2012)*, report prepared by Department of Communication and Journalism, University of Pune, submitted to the Press Council of India.

21. See also Kunhikrishnan (2006).

22. A similar process has been observed in the Hindi-speaking regions, where regional press has proliferated in the wake of the rise of satellite television. According to Sevanti Ninan, 'The rise of television and its penetration into the rural hinterland created a hunger for news. Across the Hindi belt, newspaper proprietors, circulation agents and hawkers alike assert that TV proved to be good for the newspaper business because it fuelled a curiosity that made the viewer turn to the next day's newspaper' (Ninan 2007: 15).

23. The English press in India (both print and television) is largely perceived as elitist and distanced from the everyday life of the regional citizens (Rajagopal 2009). The regional newspapers have outgrown the English press in the 1970s (Jeffrey 2000). The divide between the English- and the regional-language press has been crucial in shaping India's politics since then. For the conceptualization of a 'split public' that emerged out of the fault lines, see Rajagopal (2001).

24. Emphasis removed from the original. For a critique of the role of mass media expansion in democratization, see Nair (2003a, 2003b). The argument is that popular press in India does not have the maturity and dexterity to 'hold on to the values of democracy in the face of increasing fragmentation of the media market and convergence of technologies (in the spheres [of] information, telecommunication and entertainment)' (Nair 2003a: 4188). Media scholars have shown that the Habermasian anxiety about public sphere turning into a hollow shell due to commercialization (see the previous chapter) or the Chomskian

fear of media's role in manufacturing consent fails to take into cognizance the complexities of mediatization of politics. For a classic case of the role of media in shaping politics in India, see Rajagopal (2001). See Christy K.J. (2015), Pillai (2012), Radhakrishnan (2005), Sreekumar (2009: ch. 5), and Zacharias (2004) for analyses in the context of Kerala.

25. See a parallel argument in the context of globalization and youth culture in Kerala (Lukose 2010: 132–62). The author suggests an escalation of the contestation between two different notions of public. 'The *political* public, a conception of the public rooted within a tradition of public politics that emerged out of the colonial period, is now confronted by a privatised citizenship linked to a conception of a properly functioning *civic* public' (Lukose 2010: 140; emphasis as in the original).

26. In more recent times the participation in political public sphere is more complicated in a highly mediatized environment with the advent of television channels and social media—now information is accessed immediately through television and new media and is corroborated with a more elaborate engagement with politics on the next day through newspapers. However, this argument is tentative, and requires substantiation through ethnographic research. Some clues into how television transforms the public sphere in Kerala can be obtained from Venkiteshwaran (2014). There are also instances where public mobilization was primarily made through social media as in the case of the 'Kiss of Love' protest (2014) against moral policing.

27. F. Muller, a disciple of Hermann Gundert, was the editor of the magazine published from Thalassery by Basel Mission. It was in circulation from October 1847 to August 1851. See Vadavathoor (2001).

28. Sam (2003) suggests that *Paschimodayam* was the third periodical in the language. According to him, there was a periodical called *Malayala Panchangam* (1846) published before *Njananikshepam* and *Paschimodayam*, which are commonly believed to be the first and second news magazines in Malayalam.

29. Such an exploration is, however, outside the scope of this book. For a general account of the history of popular science literature in Malayalam, see Balakrishnan (2007).

30. The rationalist movement of Kerala that upholds 'scientific rationality' and opposes 'irrational' tendencies in society such as superstitions and religious beliefs also became active during the period. Although there had been several groups since the 1920s propagating rationalism in Kerala, it was only in 1966 that a state-level organization called Kerala Yukthivadi Sangham was formed. However, the organization became active only by 1976 (Pavanan 2002). There were also other organizations which contributed to this emerging paradigm of public engagement with science, such as the Cochin Science Society and the Chaliyar Jalavayu Shudheekarana Committee. Several local-social movements

raised environmental concerns in this period. Unfortunately, there is no comprehensive study available on these organizations and movements. Some clues to the existence of such a wide spectrum of civil society initiatives are available in Kunhikkannan (2014).

31. The formation of the Samithi in 1957 was the culmination of the discussions among a group of science writers and enthusiasts who used to meet regularly to discuss science (Namboothirippad 2014: 135). The first meeting of the Sasthra Sahithya Samithi selected Koru Master as its president and P.T. Bhaskarapanikkar as vice president. O.P. Namboothirippad was nominated as the secretary. The executive committee of the new organization consisted of Dr S. Parameswaran, C.K. Moosathu, M.C. Namboothirippad, E.V. Dev, and M.N. Subrahmanian among others (Namboothirippad 2014: 135). The Samithi published a few popular science books in the next few years (Namboothirippad 2014: 135).

32. The first meeting that decided to form the KSSP was convened on 8 April 1962 at Hotel Imperial, Kozhikode (Adiyodi 2006). The new organization had Dr K. Bhaskaran Nair as its president, Konniyur Narendranath as vice president, Dr K.G. Adiyodi as secretary, and N.V. Krishnavarier as treasurer. Lakshmi Pillayathiri Amma, Adiyodi's mother, served as the informal treasurer of the KSSP (Adiyodi 2006: 68).

33. John C. Jacob, one of the pioneers of the environmental movement in Kerala, recollects in his autobiography that he was not offered membership when the organization had begun, for he was not a science-writer (J.C. Jacob 2009: 54).

34. Many of them joined the Bhabha Atomic Research Centre (BARC) in Bombay after completing their training from the Moscow Power Institute of the erstwhile USSR.

35. Seven such associations were formed around the period. Others were Sasthra Sahithya Parishad (Telugu), Vidnyan Parishad (Marathi), Vijnjan Parishad (Kannada), Vigyan Parishad (Hindi), and Vinjana Tamil Valarchikkazhakam. A federation of these organizations also was formed, known as the Federation of Indian Language Science Associations (FILSA).

36. K.G. Adiyodi, 'Sasthra Sahithyakaranmarkku Oru Sanghatana', in *Sastra Sahitya Parishat Ulghatana Smarakam* (1962): 9, quoted in Isaac and Ekbal (1988: 5).

37. *The Social Function of Science* authored by J.D. Bernal (1939) was the best expression of the view of the Marxist radical scientists within the Social Relations of Science (SRS) movement (1932–45) in Britain. They believed that Marxism offers a methodology to connect science to society and argued for decoupling science from capitalism to be used as a vehicle for social transformation. Bernal proposed a 'science of science' to study the political economy

of science, based on Marxist methodology. The experiences of the USSR were a great inspiration for his theorization of the social relations of science. See Elzinga (1988) for more on the topic. Jawaharlal Nehru was deeply influenced by the book, and Bernal visited India several times during Nehru's tenure as the prime minister of India and gave advice to the government. Bernalism has had a strong influence on Indian science policy.

38. *Sastragathi*, the official organ of the KSSP, was launched in 1967. The movement also launched two science magazines for students: *Sastrakeralam* (1969) and *Eureka* (1970). In 1971, the KSSP launched its own independent publishing wing called STEPS (Scientific, Technical and Educational Publishing Co-operative Society), which eventually became defunct when the KSSP directly started publishing books in the mid-1970s. For more details, see Balakrishnan (2007).

39. In 1978, the KSSP had 600 village science forums. A similar initiative to start factory science forums but failed. See KSSP (1993).

40. This was also the year that marked the origin of the 'Kerala Model' of development, first proposed in a study carried out by the Centre for Development Studies (CDS), Thiruvananthapuram (CDS 1975).

41. The 1970s was a period imbued with optimism in scientific progress as we have seen earlier. This general mood of science appreciation in the early 1970s led to the creation of several S&T institutions such as the Kerala Agricultural University, Thrissur (1971), the Cochin University for Science and Technology (CUSAT), Kochi (1971), Kerala State Electronics Development Corporation Limited (KELTRON), Thiruvananthapuram (1973), Kerala Forest Research Institute (KFRI), Peechi, Thrissur (1975), Regional Research Laboratory, Thiruvananthapuram (1975), National Transportation Planning and Research Centre (NATPAC), Thiruvananthapuram (1976), Sree Chitra Tirunal Medical Centre, Thiruvananthapuram (1976), Centre for Earth Science Studies (CESS), Akkulam, Thiruvananthapuram (1978), Centre for Water Resources Development and Management (CWRDM), Kozhikode (1978), and Tropical Botanic Garden and Research Institute (TBGRI), Palode, Thiruvananthapuram (1979). The State Committee for Science, Technology and Environment (STEC) was formed in 1972. The then CM of Kerala, C. Achutha Menon, and his government (1970–7) took a keen interest in initiating the process of scientific and technological advancement of the state. The major hydro-electric power projects of the region, such as the first phase of the Idukki Hydroelectric Power Project (commissioned in 1976), also began around the time. The Idukki project had the first arch dam of India across the River Periyar.

42. The KSSP started a School for Technicians and Artisans (START) in 1976 to equip the artisans and technicians who lacked formal education, and a series of lectures on 'Nature, Science and Society' and 'the Wealth of Kerala' were

organized (*Parishad Vartha*, 1–15 October 2001). In its second activists' meet at Idukki in 1976, they decided to collect data about the natural resources and the analysis was published as a book titled *Keralathinte Sampath* (The Wealth of Kerala) (Isaac and Ekbal 1988: 45). In the same year the KSSP held a seminar on industrialization as part of its thirteenth annual meeting in Kannur. A seminar on natural resources was organized in 1977and the science caravan (Sasthra Kala Jatha, 2 October–7 November) of the year raised the slogan 'industrialize or perish'. In the next year (1978), the KSSP convened a study on the developmental crisis of Kuttanadu, the heart of paddy cultivation of Kerala.

43. It was not just the KSSP that published popular science in the 1960s and 1970s. There were several science magazines published such as *Science* (published by Puthuppalli Raghavan, 1966), *Vijnanakairali* (Kerala Bhasha Institute, 1969), *Priyarama* (1978), and *Sasthrapadham* (1990s), indicating that the trend was much widespread. See Vadavathoor (2001).

44. See Bandyopadhyay and Shiva (1988), Elzinga and Jamison (1986), Raina (1997, 2007), and Visvanathan (2007) for details.

45. According to Ramachandra Guha (1988), the emergence of the Indian environmental movement can be dated to 1973, the year of the beginning of the Chipko Andolan. See also Guha (2013).

46. See Varughese (2015b) for a detailed discussion of the emergence of different kinds of publics in India during the period.

47. These ecological groups were diverse in their approach to nature and modern science. The Zoological Club of Payyannur College started by John C. Jacob along with approximately 60 nature clubs in colleges were at the forefront of anti-Silent Valley campaign in 1977. Organizations like the Society for Environmental Education in Kerala (SEEK, 1979), Silent Valley Protection Samithi (Kozhikode), and Save Silent Valley Samithi (Thiruvananthapuram), as well as environmental magazines like *Myna* (1973), *Soochimukhi* (1981), and *Ankh* (1986) represented these groups (see J.C. Jacob 2009; James 2010; Prasad and Krishnan 2016; Radhakrishnan and Koottummel 2010; Unnikrishnan 2008).

48. The KSSP even organized a five-day study camp in early 1980 to understand and develop a critique of the epistemological debates on science (Isaac and Ekbal 1988: 43). Another important event was the visit of Ivan Illich in September 1978. He engaged with KSSP activists at the Parishad Bhavans of Kollam and Thiruvananthapuram as well as at the CDS, Thiruvananthapuram (*Parishad Vartha*, September, 1978).

49. Silent Valley is a tropical rainforest of high biodiversity in Palakkad district in northern Kerala where the state government planned to establish a hydroelectric power project in 1973. The preparations for the same started in 1976. In 1977 an agitation against the project was formed by several civil society

organizations, raising environmental concerns. The KSSP officially joined the movement only in 1979 and eventually became the face of the movement. As a result of the mass struggle the state government gave up the project in 1983. See D'monte (1985), James (2010), Radhakrishnan and Koottummel (2010), and Zachariah and Sooryamoorthy (1994).

50. The KSSP had addressed environmental issues before its participation in the Silent Valley movement. In the context of the first Stockholm conference on environment (1972), the KSSP had organized a seminar on environment in collaboration with the Cochin Science Society (Parameswaran 2002: 128). It had organized a post-Stockholm conference too (Parameswaran 2002). The organization also raised the issue of air and water pollution caused by the industries in Kochi. In 1978–9, the KSSP supported the local villagers' struggle against the issue of pollution of the Chaliyar River caused by Mavoor Gwalior Rayons, creating a debate on the effective scientific management of the issue (Parameswaran 2002). Another important endeavour was to study the ecological crisis of Kuttanadu in 1978, the wetlands of Kerala known for rice cultivation (Isaac and Ekbal 1988). For more on the KSSP's early engagements with environmental issues, see Kunhikkannan (2014).

51. In 1977, Professor M.K. Prasad presented a resolution against the dam project in Silent Valley in the KSSP's activist camp in Kaladi, but it was rejected by the engineers of the Kerala State Electricity Board (KSEB) (Radhakrishnan and Koottummel 2010: 26–7).

52. The activist camp in Kaladi (1977) appointed an expert committee to study the issue. The team consisted of Professor M.K. Prasad, Professor V.K. Damodaran, Dr K.N. Shyamasundaran Nair, Dr M.P. Parameswaran, and Dr K.P. Kannan. See Prasad et al. (1979) and Sreedharan (1990).

53. The KSSP's perspectives on the Silent Valley controversy are discussed here on the basis of articles from the period collected in Sreedharan (1990), unless mentioned otherwise.

54. The KSSP strongly opposes nuclear power plants and participated in the successful public campaign against the proposed nuclear reactor at Kothamangalam near Ernakulam (1986).

55. These projects were Idukki (third phase), Sabarigiri (augmentation), Kakkad, Kallada, and Lower Periyar.

56. Available data shows that in 1982–3 it had 566 women members out of a total membership of 5,859, that is 10.1 per cent of the total membersip. By 1988–9, this goes a little down to 9.1 per cent (3,416 out of 37,653) (see Zachariah and Sooryamoorthy 1994: 142).

57. In 1978 the KSSP took the initiative to bring together like-minded activist groups from all over the country to discuss the idea of 'people's science movement' (janakeeya sasthra prasthanam). A second convention was organized

in 1983. The proceedings of the two conventions are compiled into KSSP (1984).

58. See Jasanoff (2005) for three kinds of civic epistemologies (communitarian, consensus-seeking, and contentious) she compares in the Western context.

59. Some of these controversies were over a new kind of fever (*elippani* in Malayalam—Jacob's disease), an unknown syndrome that spoiled coconut trees (*mandari*), a lay person's claims over the discovery of green petrol (Ramar Petrol), and an alleged espionage at the ISRO, Thiruvananthapuram (the spy case), just to name a few.

60. A pilot survey that I conducted with an aim to pin down the science-related debates in two major Malayalam newspapers (*Malayala Manorama* and *Mathrubhumi*) between 1990 and 2004 clearly indicates the high frequency of scientific controversies in the media. Forty-two major scientific themes were identified as being deliberated in the newspapers over a period of 15 years. Many of these controversies were sustained in the newspapers for more than one month.

4 Public Regulation of Science
Clinical Trials at the Regional Cancer Centre

The outbreak of and the subsequent negotiations over a dispute among the scientists at one of the most prestigious medical research institutions in Kerala, the Regional Cancer Centre (RCC) in Thiruvananthapuram, which is the most important cancer research and therapy centre in the state, is the focus of this chapter.[1] The functioning of the institute and its role as a research institute came under public scrutiny in the aftermath of a scientific controversy that appeared in the regional press in July 2001. The controversy revealed the internal conflicts within the scientific institution and the ambiguities surrounding scientific knowledge production practices in biomedicine. It raised serious concerns about the ethical protocols in research, scientific collaborations with Western research institutions, changing configuration of biomedical research under neoliberalism, and patients' rights. Public anxiety about subjecting unassuming patients who seek cancer treatment at the institute to undisclosed medical experiments was central to the evolution of the controversy. It particularly offers us insights into the regulatory mechanism that came under public scrutiny. The controversy was one of the earliest of its kind that alerted the medical research community and policy makers about the escalating commercial interests in the drug and pharmaceutical sector in India.

The events that triggered off the controversy occurred between 12 November 1999 and 8 April 2000, when 27 oral cancer patients awaiting surgery were injected with chemical substances called tetra-O-methyl nor-dihydro-guairetic acid (M4N) and tetraglycinal nor-dihydro-guairetic acid (G4N) (Krishnakumar 2001). These phase I clinical trials carried out at the RCC through a contract research organization (CRO) called Quintiles India[2] were part of an ongoing drug research carried out by Dr Ru Chih C. Huang, faculty member of the Krieger School of Arts and Sciences of the Johns Hopkins University (JHU), USA (Krishnakumar 2001).[3] The clinical trials were administered by the research team at the RCC, headed by Dr M. Krishnan Nair, the then director of the institute.[4]

Dr V.N. Bhattathiri, Associate Professor of Radiotherapy at the institute,[5] filed a complaint regarding the clinical trials with the Human Rights Commission (HRC) of Kerala on 18 March 2001 (Krishnakumar 2001). The complaint 'sought a detailed inquiry, appropriate action for the violations and preventive steps', and stated that Dr Krishnan Nair, who was the 'principal investigator', should be held responsible for the unethical drug trials (Krishnakumar 2001: 124). The complainant also contended that the clinical trials had been performed before the study was sanctioned by the Drug Controller General of India (DCGI) (Krishnakumar 2001). Civil society organizations and a Malayalam television channel called Surya TV initially brought the issue to public attention. The TV channel's prime time programme called Aniyara,[6] which used to discuss issues of political and social concern with a flavour of investigative journalism, took up the issue and constantly followed it up.[7] Slowly the issue developed into a public controversy when the regional press took it up, opening an intense public deliberation. Although it remained unacknowledged by most of the regional newspapers, the role of Surya TV was crucial in exposing the issue. It should also be noted that without the intervention of the newspapers, the dispute would not have probably prompted public deliberations that continued for the next six months.

Drawing the Battle Lines

The issue was first reported in the newspapers in July 2001 when a petition was submitted to the chief minister (CM) of Kerala,[8] demanding a proper enquiry into corruption and other 'disorders' in the RCC, and a news report on the petition appeared in Madhyamam.[9] It was stated that the chemicals involved in the clinical trials had already been banned in many of the developed countries. The petition alleged that the RCC had injected a chemical substance that catalysed cancerous growth in patients without their consent and exported the malignant tissue collected from patients to the JHU for further research.[10] The petition was not reported in other newspapers, but Surya TV's reporting invited wide public attention. After a while, *Madhyamam* reported a demonstration organized by the Thiruvananthapuram district committee of the Youth Congress in front of the RCC, demanding a vigilance enquiry into the clinical trials.[11] Other newspapers continued with their silence, indicating that the battle lines were fixed well in advance, before the issue erupted into public view: Dr V.N. Bhattathiri, Surya TV, and *Madhyamam* were on one side, and the director of RCC, Dr Krishnan Nair, and the other newspapers on the other side.[12]

The divide became more noticeable in the following days. The news report in *Madhyamam* attempted to expose the ongoing conflict within the RCC over the corrupt practices of its director and his allies. The newspaper referred to the verdict of the Kerala High Court on a petition filed by Dr V.N. Bhattathiri regarding his removal from the chairpersonship of the Head and Neck Clinic.[13] The petition was filed in the wake of a major administrative reform in the RCC, when new speciality clinics were inaugurated in the institute. Interestingly, it was only *Malayala Manorama* and *Kerala Kaumudi*—the two main dailies which backed the RCC director throughout the deliberations—which reported the administrative reforms in the institute and the reports were highly appreciative of the reforms.[14] The High Court urged the RCC to settle the seniority dispute amicably within the institute.[15] After the High Court's verdict on the issue, the Citizen's Protection Forum of Thiruvananthapuram filed a public interest litigation (PIL), questioning the administrative reforms that denied the right of the patients to consult the doctor of their preference, and questioned the director's right to regularize temporary appointments.[16] The PIL also demanded cancellation of a recent extension of the tenure of the director. The allegations clearly show that the conflict within the institution was multifaceted and with a long history, finally generating public debate as a controversy over the clinical trials.

All the newspapers continued to ignore the issue even at this stage of high newsworthiness,[17] but *Madhyamam* carried on with its exposition of the malpractices in the institute. It reported the ban on the research projects of the School of Medicine of the JHU by the US government,[18]

emphasizing the ethical violations involved in the clinical trials undertaken by the RCC for the JHU. This ban was enforced following the
death of Ellen Roche, a healthy woman who was a volunteer for the
clinical trial of a drug used to treat asthma, allegedly due to the injec-
tion of an experimental drug.[19] It was reported that the state withdrew
all federal funding to the School of Medicine of the university and criticized the researchers for not informing the volunteers about the possible side effects of the drug.[20] The news report in *Madhyamam*, thus,
suggested the inhuman nature of the medical research being carried
out at the JHU and made public the research collaboration between the
university and the RCC.

These preliminary interventions of different actors that included
a scientist, a television channel, a newspaper, and a few civil society
organizations were minimally successful in pushing the issue into the
public domain. The RCC administration that had thus far been silent
was now forced to speak, and it responded to the allegations through
the newspapers which had been defending the institute through their
active silence so far. This was the moment of initiation of the scientific
public sphere. The newspapers in the following days reported intensely
on the controversy, spinning it into rigorous public deliberation and in
due course enrolling a wide range of actors. *Kerala Kaumudi* reported
the RCC administration's ardent denial of the allegations, pointing out
that it was not NDGA (nor-dihydro-guairetic acid), which had been
banned by the Food and Drug Administration (FDA) in the United
States, that was administered but M4N, a derivative of NDGA, which
was neither fatal nor banned.[21] It was also stated in the news report
that M4N was found to be completely secure in the first phase of the
experiment in different countries including the United States. The
research institute further claimed that its ethical committee had actually approved the clinical trials and consent had been obtained from all
the 25 patients to whom the drug was administered.[22] The argument
was further supported when Dr Krishnan Nair, the senior researcher
who was responsible for the clinical trial, released a press note defending the experiment.[23] Dr Krishnan Nair reiterated that the drug used
for the clinical trials was not a banned drug and that it did not have
serious side effects, but rather had a positive impact on the majority of
patients. It was also claimed that out of the patients, 16[24] were cured
and 5 responded positively, only 1 patient showed the development of

a secondary tumour. Responding to an allegation that 2 patients who had been subjected to the clinical trials died,[25] the director suggested that the drug was not the cause of death, but medical complications such as diabetes were to be held responsible (Krishnakumar 2005).[26] The careful bracketing out of the death of the patients from the reporting of the success of the experiment also should be noted here. Dr Krishnan Nair said that Dr Ru Chih C. Huang and her team would arrive on 2 August and conduct a press conference explaining the research project to the media (Krishnakumar 2005).[27]

Governmental Interventions

The dispute entered a new phase during the following days when A.K. Antony, the then CM of Kerala, entered the scenario with a damage control mission. All the five newspapers began reporting the issue from this point. The long silence of many of the newspapers on the dispute was broken by now, and the control over the reins had gone out of the hands of the backstage managers who had tried to suppress the controversy's public outbreak so far. The CM declared that his government was highly concerned about the unfortunate developments and would enquire into the charges of the malpractices in the RCC, though utmost care would be taken to safeguard the credibility of the institute.[28] The CM also hinted that punitive measures would follow if the allegations were confirmed.[29] The central government was also reported to be concerned about the issue.[30]

Such statements from the government had a very strong impact on the debate, although the state was clearly performing its conventional role of safeguarding the scientific institution from public exposure. In the following days, the RCC administration became very vocal in their denial of the allegations. It was reiterated that the drug used in the RCC was the harmless M4N[31] and not the banned NDGA.[32] It was explained that these were two different chemicals and unlike the latter, M4N did not dissolve in water and thereby did not affect other organs.[33] This explanation was to refute the allegation that the drug under experiment affected mainly the liver and the kidneys. The tumours were removed from the patients 72 hours after the drug was injected in order to conduct pathological tests, the administration explained.[34] It is important to note that the controversy was no more limited to the external 'social' aspects of the experiment; rather, the core 'technical' aspects of the experiments were mooted and debated in the scientific public sphere and the researchers were forced to defend their research by divulging out more and more scientific details.

The RCC emphasized the transparency of the clinical trials when they contended that the patients were informed about the experiment a priori and that a letter of consent was signed by all the patients and their relatives.[35] It was emphasized that treatment was provided free of cost to the patients, adding a humanitarian and ethical dimension to the clinical trials in their attempt to safeguard themselves from public ire. Such research projects had to be undertaken by the RCC since it was part of its mandate as a leading cancer research institute, the public was told.[36] It was clarified that the anti-viral agent developed by the JHU was also undergoing clinical trials in South Korea to indicate that the clinical trials in the RCC was a common research practice.[37] In India, the institute revealed, the ethical committees of the Banaras Hindu University (BHU) as well as the J.K. Cancer Institute at Kanpur had agreed to collaborate with the research project of the JHU.[38] A similar study on the impact of the drug on skin cancer was to be undertaken in Singapore.[39] Such references were used to reassure the public about the wide social acceptance of clinical trials as well as to emphasize its status as a normal and integral part of science. This move was intended to check the erosion of public trust in the institution as well as the credibility of the RCC clinical trials.

Interestingly enough, the RCC also warned that any public criticism of scientific research would affect the progress of the institute as well as the trust of the patients admitted to the institute.[40] It was pointed out that such negative criticism would discourage international research institutions and multinational drug companies from entering into research collaboration with the institute. The RCC administration was clearly attempting to situate the institute as possessing a high global reputation as a frontier research institution—any attempt to bring politics into science, hence, would be detrimental. The endeavour of the RCC administration to defend the institution was strongly supported by the RCC Employees' Association when it demanded an immediate scrapping of the public debate on the issue.[41] The Employees' Association pointed out that any public debate on the issue would only tarnish the reputation of the RCC, and hence insisted on an immediate withdrawal of all the parties from the controversy.[42]

The RCC director called for a press conference[43] in which he reiterated the earlier arguments, but ostensibly divulged more technical details about the contract research project. He claimed that the experiment had been continuing at the RCC since 1998, and the patients were chosen for clinical trials with the permission of their doctors.[44] The director pointed out that none of the doctors had opposed the clinical trials at that stage.[45] He also stated that the drug (M4N) was a derivative of NDGA developed by the JHU. The drug was proven to be harmless and effective in experiments conducted on 36 mice. Following the success of the experiment on laboratory animals, the JHU approached the RCC for undertaking clinical trials on humans, the journalists were told.[46] The RCC was approached because the drug was targeted at oral cancer, which was very common in India.[47] The RCC had received 2.5 million rupees per year for the clinical trials. The director also clarified that the RCC and the JHU would equally share the royalty of the patented drug as collaborators in the project, which later turned out to be a blatant distortion of facts.[48] Nevertheless, he offered to enquire into the allegation that another chemical substance called G4N was experimented at the RCC though the ethical committee of the institute had granted clearance solely for the trials of M4N.[49] Dr Krishnan Nair agreed that the approval from the DCGI was not received despite moving the application in September 1999.[50] Although the role of the institute was minimal in conducting the test of the drugs under trial on humans for a Western research institution, the RCC administration and its director continuously tried to confuse the public by projecting the experiment as a collaborative research with a world renowned university department in the United States.[51]

A few days later, the regional press reported that an enquiry had been ordered by Dr C.P. Thakur, the then health minister of the Government of India (GoI).[52] The minister emphasized that the enquiry would also look into whether M4N had been banned in the United States.[53] The state government in turn appointed a single-member enquiry committee (the Purvish Parikh Commission) to probe into the clinical trials controversy.[54] The HRC of Kerala accepted the complaint filed by Dr Bhattathiri and summoned the RCC director to appear before it since the RCC failed to submit an explanatory report on the clinical trials as demanded by the commission.[55] In its editorial, *Mathrubhumi* raised the legal and ethical issues associated with the controversy and demanded a judicial enquiry into the matter.[56]

The intervention of the state government was, however, not wholeheartedly welcomed by the rival coalition. Dr Bhattathiri criticized the state government for trivializing the issues he had raised by appointing Dr Purvish Parikh, who was a junior doctor, as the single-member enquiry commission.[57] Bhattathiri demanded that a committee consisting of medical and legal experts and social activists in lieu of the Parikh Commission be instituted to carry out a proper investigation into the dispute.[58] Following Dr Bhattathiri, the *Madhyamam* criticized the creation of a single-member commission, pinpointing the allegation that the objective of the commission was to safeguard the RCC director who was quite influential at the governmental level.[59] The newspaper also pointed out that Dr Krishnan Nair and Dr Purvish Parikh were friends and that they had even co-authored a research paper.[60] By challenging the technical and moral legitimacy of the expert committee, rival actors were trying hard to expose the tacit nexus between the state government and the RCC administration. They strongly opposed the governmental strategy to salvage the RCC and its researchers.

Strengthening of Coalitions

The entry of governmental regulatory mechanism stirred up the controversy, offering space to new actors who added fresh dimensions to it. The newspapers continued to be allies in the already formed coalitions, but by this time they had begun to report the opponents' arguments more openly, stripping down the veil of silence. This openness of the media indicated that the controversy had gained full momentum by then, growing beyond the control of any single actor or coalition. The deliberations during the following days indicated that the rival coalitions were gradually acquiring legitimacy and acceptance with the public. This was catalysed by the JHU's direct entry—the university published a disclaimer on its website denying any association with the clinical trials undertaken by the RCC. This denial was widely reported, and it was highlighted in the regional press that no financial assistance was offered by the university to the research project undertaken by the RCC,[61] and Dr Ru Chih C. Huang had been directed to withhold the

research.[62] The university administration constituted an enquiry committee to probe into the allegations and stated that punitive action
would be taken if the protocols and code of conduct of the university
were breached by its faculty member.[63] This act from the side of the JHU
devastatingly eroded the public support and trust enjoyed by the RCC
and its researchers. The deterioration of public trust in the institute and
the gradual public recognition of the rival scientists' arguments led to
a diversification of the subjects debated and hence expanded the scope
of the controversy.

The ethical and political aspects of the medical research were a
crucial concern in the scientific public sphere at this juncture and the
newspapers cultivated such a debate by presenting expert opinions.
Malayala Manorama on its editorial page presented the arguments of
the RCC director and the counter arguments of Dr C.R. Soman, a public
health expert with great public acclaim for his timely interventions as
a public intellectual and social activist.[64] In his article, Dr Krishnan
Nair contended that the controversy was part of an ongoing slander
campaign against him and the RCC.[65] On the contrary, Dr Soman's
arguments were strictly impersonal and technical. He pointed out that
M4N was not radically different from the banned drug, as the former
was a tetra-methyl derivative of NDGA, the basic molecule.[66] He also
argued that G4N which had been used in the clinical trials along with
M4N was a tetra-glycinal derivative of NDGA. Further, he explained
that the primary objective of the research was to test whether these
chemicals would decelerate the growth of the human immuno virus
(HIV) and later it was tested specifically in the treatment of oral cancer,
and to study its impact on the growth rate of the human papilloma-
virus (HPV). Dr Soman alleged that the protocol for medical research
was not properly followed by the JHU and the RCC as the chemical was
directly administered to human beings after conducting experiments
on mice. The drug under trial was injected into cancer patients who
were not in a critical stage of the disease, he noted. He pointed out that
this was against medical research ethics which allow such dangerous
clinical trials only on chronic patients for whom all available treatment
had been exhausted. Dr Soman contended that it was not possible to
say that the chemicals did not cause any side effects, because it was for
the first time that clinical trials were being performed anywhere. The
argument that the RCC ethical committee had approved the clinical

trials was preposterous, he pointed out, since the RCC director himself was the chairperson of the committee. He also alleged that a new drug company was registered in Singapore to manufacture the drug, if the research was successful. The claim that the RCC would get royalty for the research was rejected by him, showing that the JHU exclusively owned the patent for the drug. Based on these arguments, Dr Soman emphasized the unethical and unlawful nature of the RCC clinical trials.

The editorial page articles in *Malayala Manorama* presented the pros and cons of the controversy by providing space for the RCC director and Dr Soman under the guise of 'media neutrality', despite the newspaper's open defence of the RCC. This indicates that even though a powerful ally of the RCC in the controversy, the newspaper was forced to enter into a dialogue at this stage, relinquishing its early strategy of suppressing the dissenting actors. The entry of Dr Soman as a strong associate, who had great public reputation as a medical science expert, made the rival coalition more powerful and he became the main protagonist of the rival coalition henceforth. On the other hand, the RCC director's attempt to gain public sympathy by construing the controversy as a slander campaign against him backfired. A lot of technical information about the clinical trials had already been available in the public domain, helping other actors to construct their own arguments. The rival coalition's success at this stage was largely due to their skilful presentation of new and relevant interpretations of the case based on these resources.

The internal ambiguities of medical research became the focus of discussion as the controversy unfolded in the following days. Two public discussions on the controversy organized by the Forum for Patients' Rights (FPR)[67] and the Indian Medical Association (IMA) were widely reported in the regional press, influencing the course of negotiations in the subsequent days. In the debate on 2 August 2001 at Thiruvananthapuram organized by the FPR, a strong opponent of the clinical trials, the prevalence of similar unethical research practices in different parts of the country was noted.[68] The forum demanded a judicial enquiry into the issue.[69] Advocate Vincent Panikkulangara, who was an activist of the forum, pointed out that a 'chemical substance' and not a 'drug' was injected into the patients and explained that a chemical substance under clinical trial attained the status of 'drug' only after getting the approval of the DCGI.[70] This ontological rephrasing seemingly aimed at drawing public attention to the risk dimension involved: it was no more a drug, but a risky chemical substance that had been injected into the unassuming patients. He opined that a doctor did not have the right to prescribe a chemical substance masquerading as a drug to a patient.[71] Dr V.N. Bhattathiri reiterated his earlier statement that the Parikh Commission was simply a charade established to rescue the culprits.[72] He also denied the charge that his expose of the issue to the public would damage the institution since such a move was inevitable to safeguard 'public interest'.[73] At the meeting, Dr Gangadharan, another member of the rival group of RCC researchers, described the unethical administration of the drug under research to Mr Gopal, a patient from Tamil Nadu, without his or his doctor's consent.[74] It was alleged that the procedural lapses in the experiments thus amounted to human rights violations.[75] In general, the meeting concluded that the clinical trials were proof of the unethical practices and human rights violations prevalent in the field of medical science research in the country. The FPR seminar thus led to the consolidation of the voices of the opponents of the RCC clinical trial.

This public mood was further accentuated by the 'open debate' organized by the IMA, on 3 August 2001, titled 'Drug Trials: Right and Wrong'.[76] While the FPR seminar accorded space exclusively to the opponents of the clinical trials, the IMA seminar was inclusive in terms of participation. Therefore, the regional press extended wide news coverage despite the seminar ending in pandemonium followed by heated arguments between the rival groups. The forum resolved that transparent guidelines had to be developed for medical research and expressed its concern over the erosion of public trust in the RCC.[77]

The erosion of public credibility of the RCC, which was simultaneously a research institute and a super-speciality hospital, was pointed out by several actors. A.K. Antony, the then CM of Kerala, requested the rival groups of scientists as well as the media to abstain from any act that might damage the reputation of the RCC.[78] He pointed out that the RCC was an institution that acquired its fame as a result of years of painstaking hard work.[79] The CM cautioned the scientists against any attempt to take the issue to the streets as it would end up in the decline of the institute.[80] He also pointed out that the state government, the central government, the HRC, and the High Court had been investigating the issue and therefore a public, 'political debate' on the issue was unnecessary.[81] And it was more or less mandated by him that any destructive move against the institute would be regarded as a serious public offence.[82] The CM's intervention was in line with the state's endeavour since the beginning of the controversy to insulate the institute from public scrutiny. The statement also indicated the government's fear of the role of the mass media in exposing the political dimensions of science.

This pro-RCC statement by the CM did not go unopposed. M.N. Vijayan, a renowned Marxist intellectual of Kerala, wrote an editorial in the *Deshabhimani* magazine[83] that was quoted widely in the regional press, suggesting that the actors in the scientific public sphere largely endorsed his views. He strongly argued for the public deliberation of science for effective governance of science and denied RCC the privilege of immunity from public scrutiny. The CM's statement that justified unscientific and inhuman research in the RCC, in his opinion, went against the democratic spirit.[84] He emphasized the need for democratization of medical research while arguing that patients' consent was imperative for the conduct of such experiments on their bodies.[85] Similarly, in a statement issued by the EMS Cultural Forum, a leftist outfit, the opinion of the CM was criticized for his defence of Dr Krishnan Nair and the unethical clinical trials.[86] The debate on safeguarding scientific institutions from social audit thus was critiqued, to assert that public deliberations were inevitable for better governance of scientific institutions. Staging of public deliberation over science in mass media was hence ascertained as a greatly positive, democratic practice to be cultivated. This also indicated that the actors and newspapers were well aware of what they were doing; they defined their participation in the scientific public sphere as a highly moral action that aimed at democratizing science.

However, the pro-RCC coalition gradually recuperated from the lethargy of loss of credibility by forming a strong defence of the institution through the enrolment of more actors. The CM's open defence was a huge motivation for the supporters of the RCC and a great hue and cry was raised against the rival coalition's attempt to spoil the institute's reputation. The RCC Doctors' Association and the Employees' Association strongly supported the researchers and demanded action against Dr Bhattathiri and Dr Gangadharan for leaking the issue into the public domain, on the grounds of violation of service rules.[87] They also asked the Surya TV and rival doctors to immediately stop the smear campaign. Both the organizations declared their solidarity with the RCC director. The health minister of Kerala also criticized the attempts to tarnish the RCC, and warned against the conspiracy mounted against the institute and its 'dedicated' director.[88] Similar charges of character assassination of the director and defaming of the institute were raised by several public intellectuals.[89] Many of them demanded the suspension of the dispute till the reports of the enquiry commissions were published or made public.[90]

The involvement of the JHU in the controversy urged the regional press to report the international spillover of the dispute concerning the RCC clinical trials. It was reported that the JHU had appointed a three-member commission to investigate the allegations against its involvement in the controversial clinical trials.[91] *Madhyamam* emphasized that in the preliminary investigation undertaken by the university it was made clear that Dr Ru Chih C. Huang had not sought permission from the scrutinizing body of the university to carry out the research.[92] The regional newspapers also took note of a move in the United States to amend the export law of the country to regulate clinical trials carried out on patients from developing countries.[93] A journalistic report that had been published in *The Washington Post* titled 'The Body Hunters' on clinical trials performed by Western scientists in the developing countries had stirred up a big debate in the United States, the readers were told.[94] Newspapers informed that the concerned law was amended in the wake of this uproar, so as to prevent the export of those drugs which were still undergoing clinical trials into Afro-Asian countries.[95] After a few days, *Madhyamam* published another report quoting the reputed journal *Science*.[96] According to the newspaper, Dr Manoj Pandey, one of the members of the RCC research team, had disclosed to *Science* that the RCC had received 19,400 dollars for the clinical trials.[97] The newspaper presented it as a strong evidence against the JHU's disclaimer on its collaboration with the RCC.[98] Quoting the report in *Science*, *Madhyamam* also exposed the university's controversial tie-up with a drug company to conduct clinical trials in four more Asian countries.[99] The newspaper reported the alleged involvement of a foreign drug company called Quintiles in the RCC clinical trials.[100]

Thus, a controversy originating in the violation of ethical codes in an institution had now snowballed into a larger critique of unethical practices that characterized the world of medical research and of the global politics of knowledge production. Several civil society organizations intervened in the dispute at this juncture demanding a fresh enquiry and argued for better regulatory measures in medical research. The IMA-Kerala took note of the allegations against the RCC and *suo motu* constituted an independent committee to investigate the case.[101] The Kerala faction of the Indian Pharmaceutical Association as well as the Prathikarana Vedi[102] demanded a more democratic, non-partisan, and transparent enquiry by a committee consisting of experts from different backgrounds.[103] The KSSP and the Human Rights Protection Council of India pressed for strong action against those who conducted unethical and inhuman research at the RCC.[104] The Prathikarana Vedi conducted its own investigations into the matter and submitted a study report to the CM, which found the RCC administration guilty of not following the guidelines for clinical trials as stipulated in the Drugs and Cosmetic Rule of 1988.[105] The report noted that the ethical committee in the RCC was not constituted in compliance with the accepted standards. The forum also alleged that the Helsinki Declaration which safeguarded patients' rights was violated by the RCC.[106] Seminars were organized on the controversy by several organizations, deliberating the pros and cons of the issue.[107] Most of them discussed the controversy in the light of the corrupt practices that afflicted medical research.[108]

The internal dynamics of medical science research, which is usually insulated from public view, thus became an open subject of public scrutiny in the context of the clinical trials controversy. The endeavour to use people of the developing countries as guinea pigs and the operations of multinational drug companies were emphasized as grave concerns for the public at this juncture, thanks to the persistent effort of the Malayalam press to report to its readers the debates taking place in the Western media. The linking of regional concerns with the debate in the West facilitated the expansion of the scope of the regional deliberations and an intense involvement of civil society organizations. There was also strong resentment about the regulatory mechanism that was introduced by the state. The governmental enquiry committees were seen as highly partial and inefficient, and civil society organizations constituted their own expert committees and research teams, which sought to demonstrate what a robust regulatory mechanism was supposed to be. These committees included a wider range of experts.[109]

The patients who had been subjected to the drug trials momentarily appeared as actors in the public deliberations at this juncture, as a result of the organized effort of the rival coalition. Dr Gangadharan, advocate Vincent Panikkulangara, and the *Madhyamam* had a key role in bringing in the voices of the patients. *Madhyamam* published an exclusive report on the victimization of the patients, presenting the case of Gopal, who was suffering from cancer of the tongue.[110] Gopal was quoted as suggesting that he had been unaware of the clinical trials. He alleged that when the drug under experiment was injected he had been misinformed that it was a local anaesthetic administered to remove the malignant tissue from his tongue. The news report pointed out that the document which explained the nature of the experiment that Gopal had to undergo was in English and Malayalam, and therefore he could not follow its contents, as he knew only Tamil. The drug did not produce any positive change in him and he came back again to the RCC with throat cancer, according to the news report. It was hinted in the news report that this could have been caused by the experimental administration of the drug. Gopal soon filed a formal complaint against the research team with the HRC of Kerala with the aid of his advocate, Vincent Panikkulangara.[111] Dr Gangadharan also appeared before the commission and alleged that the drug had been administered to his patient without his consent.[112]

Later, advocate Panikkulangara appeared before the Parikh Commission and stated that two of the patients who had been administered the drug died within 50 days, and submitted the case documents from the RCC as evidence.[113] Lalithamma was one of the female patients who allegedly died after undergoing the clinical trials; her son too approached the enquiry commission appointed by the central government and complained that his mother passed away after being subjected to the experiments.[114] Thus, 'the suffering of survivors became an occasion for the exercise of power' in the scientific public sphere (Das 1995: 156). The RCC director denied the charges forwarded on behalf of the patients pointing out that the deceased was a chronic cancer patient at an advanced stage of illness and therefore her death was not caused by the drugs under trial.[115] On the day when Gopal's story appeared in *Madhyamam*, a joint meeting of the staff associations of the

RCC alleged that the rival scientists were trying to provoke people by propagating that the clinical trials were carried out on Dalit patients.[116] They also criticized the media for selectively presenting the case of terminal patients before the public.[117]

The rival coalition's strategy to employ the discursive frame of patients' rights eventually lost ground, apparently because of the lack of convincing medical evidence to substantiate the claim that the patients developed secondary tumours due to the drugs under trial. The suffering of the patients who underwent the experiments was thus delegitimized through the effective employment of a medico-legal discourse (Das 1995). Although the attempt to raise the question of patients' rights largely failed, the image of the patient's body as subjected to the unethical and harmful drug trials by the medical research establishment had been actively maintained in the deliberations, thereby separating the real victims from their experience of suffering. As Veena Das has argued in the context of the Bhopal disaster, 'every reference to victims and their suffering only served to reify "suffering" while dissolving the real victims in order that they could be reconstituted into nothing more than verbal subjects' (Das 1995: 164). Since the RCC had been the major medical institution available in the region for cancer diagnosis and treatment, the controversy over unethical research practices was perceived in connection with the risks that the global politics of medical science research created for patients in developing countries like India. References in the public discourse to the patient's violated body reinforced the perception of the possibility of being subjected to dangerous chemical substances under the guise of cancer treatment, while the patients who underwent the clinical trials themselves lost their voices in the scientific public sphere.[118]

'Safeguard the RCC!' The Public Consensus

However, a general consensus about safeguarding the status of the RCC as a reputed medical research institution gradually began to emerge in the scientific public sphere. At the same time, it is evident that the crystallization of this consensus did not close the controversy; deliberations continued, but most of the actors agreed on defending the institution while continuing to debate the issues involved. This consensus was accomplished through the concerted campaigning of a wide range of pro-RCC actors. Although the need to safeguard the institution had begun surfacing since the CM's plea to the public, it was the public intellectuals and literati of Kerala who forced it as a central concern for all Keralites. They signed a joint open letter that requested the public to desist from tarnishing the institution while criticizing the clinical trials.[119] Many of the intellectuals used the 'Letters to the Editor' columns as well to express their concerns.[120] The 'RCC Protection Forum' was a general platform created exclusively for the purpose of the campaign by the pro-RCC actors, mainly the staff of the institute. They had rigorously campaigned against any attack on the RCC through demonstrations, public meetings, and press conferences.[121] There were also a considerable number of letters to the editor that appeared in different newspapers authored by medical professionals.[122] Most of them found the allegations against the RCC clinical trials as a threat to the autonomy of the medical establishment itself, and thereby to their own professional authority. These interventions clearly indicate how professional interests motivated actors to participate in a controversy.[123] It is also interesting to note that while the entire RCC staff and their organizations defended the institution, most of the youth wings of political parties (especially the All India Youth Federation [AIYF], the Democratic Youth Federation of India [DYFI], and the Youth Congress) came out against the RCC and demanded strong action against the culprits.

The *Kerala Kaumudi* was a major actor that supported the campaigners by relentlessly voicing their concerns and reporting their campaigns. The newspaper strongly defended the RCC and its director in its editorial on 18 August.[124] The editorial argued that Dr Krishnan Nair must not be criticized, as he was the singular person who created the institution from almost 'nothing'.[125] The 'Letters to the Editor' column was meticulously utilized by *Kerala Kaumudi* to articulate the campaigners' concerns.[126] The newspaper skilfully selected and published letters to the editor, a strategy which helped the newspaper in voicing its own concerns as readers' opinion. This was accomplished mainly by organizing the column in such a way that the majority of letters expressed the views and interests of the newspaper, with a few others presenting counter-arguments. Most of the letters that appeared in the newspaper expressed the readers' reservations about the orchestrated attempt at character assassination of the RCC director, which was seen as a serious threat to the reputation of and public trust on the institute.[127] Some of the letters supported the statement of the CM and in so doing defended the institute and the director.[128] Many readers attacked the rival group of scientists for their ulterior motives.[129]

Some of the letters which presented the counter argument linked the controversy to the wider question of ethics in medical science research. One reader, for example, invited public attention to the fact that several of the drugs banned in the West were still used in India.[130] This unethical evasion of all regulatory measures was indicative of the larger malaise exemplified by the RCC clinical trials, the letter argued. Another reader, a doctor by training, expressed the same concern by pointing to the unholy alliance between doctors and multinational drug companies.[131] As a solution, he suggested medical research be restricted to universities and demanded the proscription of all contract research projects being carried out for foreign agencies. He also suggested the formulation of a new drug policy that curtailed unethical and inhuman practices in medical science through stricter regulatory measures. However, the genre of letters to the editor was not intensely used in this manner by other newspapers in the clinical trials controversy, despite a large number of letters being published.

Such an organized campaign had a deep impact on the course of the dispute. Nevertheless, by then the need to safeguard the RCC was disconnected from the democratic imperative of constant public scrutiny of scientific institutions. The consensus shared by the participants in the deliberations (irrespective of their affiliations) was that the prestige and reputation of RCC as a medical institution had to be safeguarded while public debate was expected to continue in reference to the malpractices in medical science research. Even the opponents of the clinical trials softened their attack against the RCC as a regional scientific institution. This is quite evident from a later statement of Dr C.R. Soman wherein he stated that he had not had any intention *to malign the institution,* and his criticism was directed against *the unethical clinical trials.*[132] The letter to the editor from a reader who identified himself as a medical doctor was highly suggestive of the mood that prevailed; he defended the RCC while strongly recommending an immediate moratorium on the controversy pointing out that an enquiry into the matter was already on.[133] In the same vein, he acknowledged the prevalence of unethical practices in medical sciences and demanded punitive action against those who prove to be guilty. Following this reorientation of the debate away from the institute, the rival alliance tried their level best to bring the RCC back to focus, but in vain. The consensus thus achieved at this stage was highly nuanced; on the one hand it was agreed that the institute had been a great medical institution that delivers meticulous service to the public, but on the other, the public anxiety about the malpractices in the institute continued, and corrective measures were sought through governmental interventions. Safeguarding the RCC did not hinder endorsing the need for continuous public scrutiny of scientific institutions in the region, and the role of mass media in creating a deliberative space for public regulation of science was ascertained, especially in the context of lack of a strong regulatory mechanism in the country to control clinical trials. The RCC case was rather influential on the future development of strict guidelines for contract research in the country, as we are going to see later.

Enquiry Commission Reports and the Closure of the Controversy

A closure of the controversy was attained in the following months as a result of fresh developments after the appointment of the enquiry commissions. Four such enquiries were on at the time: the Parikh Commission constituted by the state government, the commission sent by the central government,[134] the HRC, and the enquiry instituted by the IMA-Kerala. The investigations of these bodies became the focal point of attention in the regional press in the last phase of the controversy, and the deliberations reflected the public demand for better institutional mechanisms for regulating science in the country. Therefore the expert commissions appointed by various governmental regulatory agencies were closely watched and their actions debated in the scientific public sphere.[135]

The Parikh Commission appointed by the state government commenced its hearing on 9 August 2001. The Commission never enjoyed great public legitimacy due to the general perception about its allegiance to the RCC administration, and the newspapers in the rival coalition tried to surf on the public sentiments. *Madhyamam* reported that Dr Parikh arrived for the hearing in a vehicle provided by the RCC.[136] The news report also claimed that the person who assisted the commission was an employee of the institute. In yet another report, the newspaper argued that the assistant originally belonged to the personal staff of the RCC director.[137] The Parikh Commission was a charade staged by the state government hand-in-glove with the RCC administration to salvage the culprits, the newspaper alleged.[138] Dr Bhattathiri declared that he would not cooperate with the commission and demanded the replacing of the single-member commission with an expert commission comprised of experts from different fields.[139] This argument gained wide currency as organizations like the KSSP and the AIYF supported the demand.[140] Dr Parikh denied all charges against the credibility of the commission and assured a just and transparent investigation.[141] The HRC was not criticized by the anti-RCC group, because its functioning was indirectly supportive of their cause from the beginning. In the days that followed, the regional press began reporting the functioning of the Parikh Commission, the Health Ministry Commission and the HRC on a regular basis. The newspapers reported the statements made before the commissions by a wide range of people who were involved in the controversy.

It was at this point that *Kerala Kaumudi* attempted to steer away the whole controversy in another direction in favour of the RCC, sensing a gradual emergence of the institute once again as the target of public criticism. The investigative news report of P.V. Murukan presented certain 'evidences' for the vested interests of the rival group of scientists in fanning the controversy.[142] According to the report, Dr Bhattathiri and Dr Gangadharan, two scientists who sparked off the controversy, had registered a cooperative society with the intention to start a cancer treatment and research institute at Kollam.[143] The report alleged that the controversy was a vicious and concerted endeavour to create public distrust in the RCC in order to promote the new hospital and research institute the rival scientists were planning to establish. The report appeared as a piece of investigative journalism; however, it was not an exclusive one. *Malayala Manorama*, another strong defender of the RCC, on the same day reported sealing of the documents of registration of the 'Janatha Health and Radiation Research Society' by police on a request from Dr Krishnan Nair.[144] The appearance of the news reports on the same day in two newspapers which had offered relentless support to the RCC director and his team of researchers suggest that this was a well-planned ambush on the rivals. The registration of the society was apparently an already known fact among the scientists and staff of the RCC, but revealed to the public at this decisive moment to win over the rival coalition.[145]

The investigative report of *Kerala Kaumudi* had a big impact. The pro-RCC speakers effectively manoeuvred the situation to their advantage:

the RCC Protection Forum attacked their opponents virulently and demanded immediate dismissal of those who were involved in the venture.[146] At this juncture, *Madhyamam* endeavoured without much success to arrest the setback by publishing an exclusive report on the financial side of the clinical trials. According to the report, the institute had received 4 million rupees from the JHU as remuneration, and the money was kept in a special bank account directly controlled by the RCC director.[147] The news report alleged that this was indicative of the financial corruption involved in carrying out the research. The newspaper also revealed that foreign funds received by the institute for the purpose of research were not audited for several years, further reinforcing charges of financial corruption on the institute.

The endeavours of the rival scientists and their allies to fight back further proved unsuccessful when the investigative reports of the enquiry commissions were released.[148] All the reports shared the consensus that had already been attained in the scientific public sphere—that the credence of the RCC as a medical research institution must not be tampered with. At the same time all the three commissions recommended revision of the medical ethics guidelines to avoid violation of research protocols in future and reconstitution of institutional regulatory mechanisms. Once the commission of the Ministry of Health and Family Welfare submitted its enquiry report, the ministry promptly suspended all the ongoing human drug trials at the RCC for six months.[149] The enquiry report revealed that the guidelines of the Indian Council of Medical Research (ICMR) for research were not followed by the RCC during the clinical trials.[150] According to the enquiry report, the research had been carried out in collaboration with the JHU and the drug involved was a derivative of NDGA.[151] The enquiry commission also revealed that research had commenced well before permission was granted by the DCGI.[152] However, the report emphasized that there were no human rights violations involved in the clinical trials and instead highlighted the infringement of ethical protocols.[153] The report pointed out that the specimens collected from the patients had been sent to the JHU without permission from the concerned regulatory bodies.[154] On the basis of the report, the Health Ministry sent a show-cause notice to the RCC and directed the institute to reconstitute its ethics committee.[155] The government decided to make the ICMR guidelines for medical research mandatory for all medical research institutions in the country.[156] It was reported in the media that punitive action against the accused would be sought by the government.[157] However, the Health Ministry of India did not feel the need for any punitive action against the institute and its researchers on the grounds that the drug used for the clinical trials was not banned.[158] The IMA-Kerala also published its report around this time, with almost the same conclusions.[159] The rival group of researchers' act of raising the matter in the media rather than 'handling it officially' was staunchly criticized in the report. Both the reports were criticized by the newspapers in the rival coalition without much impact. They contended that by rejecting the charge of human rights violations in the clinical trials, the reports safeguarded the scientists and the RCC administration against any strong punitive action.[160] Dr V.N. Bhattathiri challenged the conclusion of the reports that there was no human rights violation involved.[161]

The Parikh Commission's preliminary report was made public on 4 October 2001, and it was more sympathetic to the RCC than the other commissions.[162] The report observed that other than 'minor procedural errors', the clinical trials were not erroneous at all.[163] According to the report the research agreement between the RCC and the JHU was transparent.[164] The commission pointed out that the drug used was not banned and it was effective when administered to patients and no side effects were reported.[165] It was asserted that prior consent had been given by the patients.[166] Similar to the IMA report, the Parikh Commission's report also criticized the rival scientists for creating a public controversy over the issue by 'misleading' the media.[167] The commission, nonetheless, suggested reconstitution of the institutional ethics committee of the RCC. The Parikh Commission report also invited much criticism from the allies of the rival scientists.[168] In a news report, *Madhyamam* asserted that the reports of the Parikh Commission and commission appointed by the Health Ministry were orchestrated attempts to absolve the RCC scientists.[169] Meanwhile, *Kerala Kaumudi* published an editorial page article and a news report strongly supporting the findings of the commissions and criticizing the tendency to pursue the controversy even after the enquiry reports had been published.[170]

Quite contrary to the enquiry reports from India, the conclusions of the JHU Enquiry Committee were against the principal researcher, Dr Ru Chih C. Huang. The enquiry report published by the JHU revealed that the researcher had not followed the guidelines of the university as well as the scientific protocols for clinical trials.[171] The JHU enquiry committee also assessed that the clinical trials undertaken by the RCC as a client of Quintiles India had several shortcomings. Based on the enquiry report, the university forbade Dr Huang from being the principal investigator in human drug trials. The stark contrast between the Indian and the US reports was highlighted by some actors, pointing to the failure of the regulatory mechanism in India.[172]

Since the publication of the reports, the regional and central governments began vociferously defending the RCC and the clinical trials. The union health minister, Dr C.P. Thakur, told the Indian Parliament that the banned drug had not been experimented at the RCC and therefore human rights were not violated.[173] The state minister of health, P. Sankaran, declared the same in the state legislative assembly, though the opposition staunchly criticized the government and the Parikh Commission.[174] The opposition demanded immediate scrapping of the Parikh Commission report for its contradictory and counterfactual findings and urged the government to order a fresh enquiry.[175] However, the health minister of Kerala reiterated that any false propaganda against the RCC would affect its reputation, although the government was willing to enquire into any complaint against the institution.[176] Based on the assessment of a cabinet sub-committee, the report of the Parikh Commission was accepted by the government in February 2002.[177]

The controversy made the public as well as policy makers realize the political, ethical, and legal complications in allowing Western research institutions to carry out clinical trials in India through CROs. On the one hand, the use of patients from India as 'guinea pigs' for drug research in prestigious Western scientific institutions became exposed and the commercial and political interests that played out in outsourcing clinical trials to client organizations like the RCC became a serious issue for deliberation for the regional scientific-citizen public.[178] On the other hand, the controversy brought to light the inadequacy of Indian regulatory measures and the lack of a proper understanding—even among Indian researchers—about the global dynamics of drug research shaped by neoliberal market economy. Following the controversy, the Ministry of Health and Family Welfare, as we have seen, took initiatives to develop better institutional mechanisms to regulate clinical research. Agencies like the ICMR also became proactive in reframing ethical guidelines for government research institutions to take part in drug trials in the country. Of late, the courts also became actively involved in regulating drug research in India (Prasad 2009). The institutional ethics committee of the RCC was reconstituted in the wake of the controversy and new ethical procedures are being set up. The RCC clinical trials and the public deliberations that followed were a milestone in the path towards developing robust ethical and legal regulatory protocols to curb the onslaught of commercial interests that eclipsed the character of science as public knowledge.

***

The public deliberations over the drug trials at the RCC brought out the internal ambivalences and complex power relations of biomedicine to the public domain. The deliberations in the scientific public sphere revealed the transformation of science within the neoliberal paradigm: in the post liberalization phase, drug research has been increasingly carried out through a global research network that brought in new private players like CROs and their client organizations like the RCC into picture. The emergence of CROs as major players in drug research represents the accelerating commercialization and privatization of scientific research (Mirowski and Van Horn 2005). While the state and the regulatory institutions were largely lackadaisical in responding to the changed paradigm, it was the mass media that brought the issue into public attention.

The controversy disclosed several unique features of the case. The deliberations revealed an important aspect of medialization of science: Originally the case had begun as a scientific controversy within the RCC when some dissenting voices emerged among the scientists about the everyday governance of the institution. Once the splinter group of scientists realized that they were vulnerable within the institute, the issue was leaked to the public with the help of the media. They actively made alliances with media institutions, journalists, and civil society actors to fortify their position. The development of the intra-institutional conflict into an intense public controversy forced the research team that conducted the clinical trials also to form their own coalitions to participate in public deliberations and defend their actions. The researchers on both sides of the dispute actively approached the media and invited public attention to gain trust and credibility. Press conferences were organized, press notes were released, and the researchers gave interviews to journalists and wrote articles in newspapers during the controversy. Suddenly, scientists found it important to engage with the media to safeguard their interests.

The regional press was not active in the initial phase to bring the dispute into the public domain. As we have seen, the issue was taken up by a news channel and some civil society organizations. However, thereafter the regional press became proactive and turned into the major site of deliberations. This vividly suggests that more than any other sites/social agents, the regional press held the potential to generate public controversies. Once the scientific public sphere became activated, new actors appeared on the front stage, and political alliances were formed. These alliances were also active at the backstage, aggressively trying to steer the controversy. They staged the performances and designed rhetorical strategies to influence the deliberations. However, due to the presence of a large number of actors, and their complex interactions, the scientific public sphere continued to enjoy deliberative autonomy, making it functionally open and dynamic. The KSSP was not an important mediator anymore, nor was it an actor of any significance in the scientific public sphere constituted by the regional press, unlike the public debates of the 1980s.[179] Nonetheless, there existed many parallel sites of negotiations in the form of citizen forums, political organizations, and other civil society groups who took serious interest in the controversy. These alternative sites of deliberation were weak and dispersed and hence highly media-dependent for organizing their campaigns. The civil society organizations and their spokespersons turned to be actors in the scientific public sphere. Their voices were amplified to a wider audience by the regional press. Any decline in media attention directly affected their visibility and status as actors. To put it differently, it was their *mediation* that shaped them into important actors in the controversy. At the same time, their participation helped expanding the contours of the controversy as well. The citizen forums and initiatives which critically participated in the controversy were successful in bringing in new perspectives and adding new dimensions to the deliberation.

Another aspect of the controversy was the perceptible existence of a strong filtering mechanism that largely determined the course of the deliberations. The regional press in the beginning of the controversy attempted to regulate the dispute in an explicit manner and many of the newspapers were strong allies in the competing coalitions. The dynamics of the journalistic production of news—the backstage activities—were strong in the clinical trials controversy. However, this does not imply that the trajectory of the controversy was engineered by a hidden network functioning behind the scene. On the contrary, the controversy demonstrated that its course of evolution had been guided by a complex deliberative process which was not controlled by any single actor/alliance. The media was not simply reflecting the public mood, but actively shaping the controversy by participating in it. The media institutions were functioning simultaneously as gatekeepers and agenda brokers. This double role of the newspapers was crucial in the initiation, evolution, and closure of the controversy.

The RCC controversy also revealed the pertinent role played by the scientific public sphere in influencing policy institutions through deliberations. The scientific-citizen public were aware of their role in democratizing science and their deliberations initiated a policy discourse on the urgency of developing new regulatory mechanisms directed towards addressing the challenges raised by the neoliberal biomedical paradigm. The scientific public sphere was more rigorous and effective in identifying and negotiating the new mode of knowledge production than the policy institutions and regulatory organizations. The concerted endeavour of the RCC researchers and their associates to project the clinical trials as frontier research they had carried out in collaboration with the JHU was foiled in the scientific public sphere where other actors always showed great clarity in understanding the internal differentiations of science. The need to develop more inclusive mechanisms for democratic governance of science and the need for a wider definition of expertise were debated and demonstrated in the public deliberations, whereas the existing regulatory structures were more recalcitrant. The commissions which looked into the case were appointed by various governmental regulatory agencies, and they spent more energy on defending the researchers who conducted the clinical trials than openly engaging with the issues and concerns being raised in the scientific public sphere. These commissions hence enjoyed less public trust, and the experts' credibility was continuously being contested in public. The emergence of the scientific public sphere was crucial in articulating the anxieties and uncertainties and enunciated new possibilities for democratic governance of science.

Notes

1. The RCC was established in the year 1981, and the institute is jointly sponsored by the central and the state governments. Several international agencies including the World Health Organization (WHO), the International Agency for Research on Cancer (IARC), and the United Nations Population Fund (UNFPA) have acknowledged the institute as a centre of excellence. See State Planning Board (2006).

2. Quintiles India (or Quintiles Spectral Ltd) was launched in 1997 as a collaborative venture between Quintiles Transnational Corporation and an India-based healthcare firm called Core Healthcare Limited (Prasad 2009). The RCC clinical trials were probably one of their initial projects. See Mirowski and Van Horn (2005) for a detailed analysis of how the rapid rise of the CROs provides impetus to the commercialization of biomedical research.

3. The Food and Drug Administration (FDA) is the federal agency in the United States which regulates drug development research. The procedures set by the FDA eventually became the basis for research guidelines all over the developed world (Mirowski and Van Horn 2005: 508). Within the FDA framework, drug research is organized into four stages. These are: a pre-clinical stage where experiments are carried out on laboratory animals, followed by a clinical stage, a regulatory delay, and a post-clinical stage (Mirowski and Van Horn 2005). The clinical stage of drug research consists of four phases where the drug under trial is administered on humans (Mirowski and Van Horn 2005). BioCure Medicals is a company that was established in 2000 by BioCure Technologies to develop M4N and other drugs. The JHU has shares in the company which is currently known as Erimos. See Krishnakumar (2005). The Phase I trials at the RCC were apparently outsourced to the Quintiles International by Erimos.

4. The research team had Dr Manoj Pandey and Dr Radhakrishna Pillai as co-researchers. See Krishnakumar (2001). As its founder director, Dr Krishnan Nair had great contribution in building the RCC since its inception (1981–2003). For an autobiographical narration on his close relationship with the institute, see Nair (2013).

5. Dr Bhattathiri had finished his PhD research on radiotherapy with Dr Krishnan Nair. See Nair (2013: 159).

6. 'Aniyara' literally means 'backstage'. The programme was directed and anchored by Alias John, a reputed social activist-turned-TV journalist, and was produced by his own company called National Television (NTV).

7. Many of the speakers interviewed by me pointed out the role of Surya TV in exposing the issue. Later on it was alleged that the TV channel had deliberately been trying to defame the RCC, and that it was hand in glove with the Adayar Cancer Institute, Tamil Nadu, though the allegation did not get much currency in the scientific public sphere. Personal interview with Dr Joy Elamon, 24 April 2006, and Dr C.R. Soman, 26 April 2006.

8. A previously unheard of organization called the All India Medical Service Centre (AIMSC) was the petitioner. The name of the organization appeared only in this particular news report. The CM of Kerala is the chairperson of the RCC governing body, and hence the petition was submitted to him.

9. 'Cancer Centreile Kramakkedukale Kurichu Anweshikkanam', *Madhyamam*, 3 July 2001, p. 2.

10. 'Cancer Centreile Kramakkedukale', *Madhyamam*.

11. 'RCCkku Munpil Dharna Nadathi', *Madhyamam*, 13 July 2001, p. 3.

12. *Kerala Kaumudi* and *Malayala Manorama* openly defended the RCC director throughout the controversy as will be discussed. *Mathrubhumi* did not explicitly support any of the coalitions, but a majority of its news reports represented the pro-RCC actors in the scientific public sphere. *Deshabhimani* took a non-partisan view on the matter, but critiqued the governmental interventions, especially in the final phase of the controversy. In my personal interview with a reputed media analyst from Kerala (name withheld), it was pointed out that many of the newspaper authorities had a cordial relationship with the RCC administration and especially with Dr Krishnan Nair. If the newspaper owners or the reputed journalists recommended a patient, he/she was given great priority and utmost care in the RCC. In return the newspapers always helped the institute maintain good public relations by publishing reports whenever needed. He also pointed out that the chief editor of one of the major newspapers was a member of the governing council of the institute.

13. 'RCCkku Munpil Dharna', *Madhyamam*.

14. Each speciality clinic was meant for the treatment of the cancer that affects a specific part of the body. A speciality clinic consists of a team of experts led by a senior doctor. See 'RCCyil Speciality Clinikkukal Erppeduthi', *Malayala Manorama*, 8 July 2001, p. 4; 'RCCyil Speciality Clinikkukal', *Kerala Kaumudi*, 8 July 2001, p. 10.

15. 'RCCkku Munpil Dharna', *Madhyamam*.

16. 'RCCkku Munpil Dharna', *Madhyamam*.

17. The one and only exception was a small news report in the local edition of *Kerala Kaumudi* about the PIL filed by the Citizens' Forum. See 'RCC Directorude Kalaavadhi Neettiyathinu Ethiraya Harjiyil Notice', *Kerala Kaumudi*, 25 July 2001, p. 3.

18. 'John Hopkinsinte Pareekshanangalkku America Vilakkerppeduthi', *Madhyamam*, 23 July 2001, p. 5.

19. The drug was 'Hexamethonium' used in combination with sodium bicarbonate, which was not scientifically allowed, according to the *Madhyamam* report. See 'John Hopkinsinte Pareekshanangalkku', *Madhyamam*.

20. 'John Hopkinsinte Pareekshanangalkku', *Madhyamam*.

21. 'Nirodhicha Marunnu Kuththivechittilla: RCC', *Kerala Kaumudi*, 15 July 2001, p. 10. The news report was quoting an explanatory note by the RCC administration to the health minister of the state.

22. 'Nirodhicha Marunnu Kuththivechittilla', *Kerala Kaumudi*.

23. His statement appeared on the next day in *Malayala Manorama*. 'Nirodhitha Marunnu RCCyil Upayogichittilla: Krishnan Nair', *Malayala Manorama*, 21 July 2001, p. 12.

24. Note the confusion regarding the number of patients. According to the news report in *Kerala Kaumudi* the number was 25, and in the report by *Malayala Manorama* the number was reduced to 22. Later in an interview with Dr Ru Chih C. Huang, it emerged that 27 patients had been administered the drug. See Krishnakumar (2005).

25. The source of the allegation is not clear. As we are going to see, the same allegation reappeared in the scientific public sphere at a later phase.

26. However, it was not discussed if the diabetes could have been drug-induced.

27. This has turned out to be a false claim.

28. 'RCC Prasnam Sarkar Thalathil Anweshikkunnu: Antony', *Malayala Manorama*, 26 July 2001, p. 9; 'RCCkkethiraya Aropanangal Anweshikkunnu: Mukhyamanthri', *Mathrubhumi*, 26 July 2001, p. 5; 'Cancer Centreinte Visvasyatha Nashtappeduthilla', *Madhyamam*, 26 July 2001, p. 5; 'RCC: Aropanangal Anweshikkum Ennu Mukhyamanthri', *Kerala Kaumudi*, 26 July 2001, p. 11.

29. 'RCC: Aropanangal Anweshikkum', *Kerala Kaumudi*.

30. 'RCCkkethiraya Aropanangal', *Mathrubhumi*; 'RCC Vivadam: Kendram Vivaram Aarayunnu', *Mathrubhumi*, 26 July 2001, p. 5.

31. The RCC explained that the first phase of the experiment on animals proved that the drug had no side effects. See 'Nirodhikkappetta Marunnu Upayogichittilla: Krishnan Nair', *Deshabhimani*, 28 July 2001, p. 8.

32. 'Nirodhitha Marunnu Pareekshichittillennu RCC', *Malayala Manorama*, 28 July 2001, p. 9; 'Nirodhikkappetta Marunnu Upayogichittilla', *Deshabhimani*;

'Gaveshana Sthapanam Ayathinal Marunnu Pareekshiche Theeroo: Cancer Centre', *Madhyamam*, 28 July 2001, p. 5; 'Cancer Chikilsaykku Upayogichathu Paarsva Phalangalillennu Thelinja Marunnu: Dr. M. Krishnan Nair', *Kerala Kaumudi*, 28 July 2001, p. 12.

33. 'Cancer Chikilsaykku Upayogichathu', *Kerala Kaumudi*.

34. 'Nirodhitha Marunnu Pareekshichittillennu', *Malayala Manorama*; 'Cancer Chikilsaykku Upayogichathu', *Kerala Kaumudi*.

35. 'Cancer Chikilsaykku Upayogichathu', *Kerala Kaumudi*.

36. 'Gaveshana Sthapanam Ayathinal', *Madhyamam*. This claim was problematic, as the RCC was just functioning as a CRO for the JHU researcher.

37. 'Nirodhitha Marunnu Pareekshichittillennu', *Malayala Manorama*; 'Gaveshana Sthapanam Ayathinal', *Madhyamam*.

38. 'Gaveshana Sthapanam Ayathinal', *Madhyamam*. See also 'Nirodhicha Marunnukal RCCyil Upayogichilla: Director', *Deshabhimani*, 29 July 2001, p. 4.

39. 'Gaveshana Sthapanam Ayathinal', *Madhyamam*.

40. 'Gaveshana Sthapanam Ayathinal', *Madhyamam*.

41. 'RCCyude Salpperinu Kalankam Undakkaruthu', *Kerala Kaumudi*, 27 July 2001, p. 3. The Post Graduate Students' Association and the Junior Doctors' Association of the Medical College, Thiruvananthapuram, expressed similar concerns. See 'RCC Vivadam: Asanka Pariharikkanam', *Malayala Manorama*, 31 July 2001, p. 9; 'Cancer Centre: Asanka Pariharikkanam', *Mathrubhumi*, 31 July 2001, p. 3.

42. 'RCCyude Salpperinu Kalankam', *Kerala Kaumudi*.

43. The press conference itself was controversial, as it had been postponed to the next day at the last moment without prior notice. According to *Madhyamam*, the public relations officer (PRO) of the RCC explained to the journalists present at the press conference on Friday, 27 July 2001, that the press conference was postponed to the next day and then distributed a press note, instead of answering the questions raised by the journalists regarding the controversy. The newspaper reported that the journalists protested and then the PRO explained that the press conference was postponed to ensure the presence of the representatives of the JHU. See 'RCCyude Pathra Sammelanam Vivadamayi', *Madhyamam*, 28 July 2001, p. 5. See Chapter 7 for a discussion on the role of press conference as a deliberative space and source of information.

44. 'Marunnu Pareekshanam Nadannathu Niyama Vidheyamayi: Dr. Krishnan Nair', *Mathrubhumi*, 29 July 2001, p. 5; 'Nirodhicha Marunnukal RCCyil', *Deshabhimani*.

45. 'Nirodhicha Marunnukal RCCyil', *Deshabhimani*.

46. 'Marunnu Pareekshanam Nadannathu', *Mathrubhumi*.

47. 'Nirodhicha Marunnukal RCCyil', *Deshabhimani*, p. 4; 'Marunnu Pareekshichathu Ethical Committeeyude Anuvadathode: RCC Director', *Malayala Manorama*, 29 July 2001, p. 7.

48. 'Marunnu Gaveshanam RCCyude Joli: Vivada Marunnu Phalaprada-mennu Thelinju: Dr. M. Krishnan Nair', *Kerala Kaumudi*, 29 July 2001, p. 1. In an interview that appeared long after the closure of the controversy, Dr Ru Chih C. Huang strongly denied this. The RCC was just a CRO who conducted the first human trials of the drug, she affirmed. See Krishnakumar (2005).

49. 'Marunnu Pareekshanam Nadannathu', *Mathrubhumi*. See also 'Maru-nnu Gaveshanam RCCyude', *Kerala Kaumudi*.

50. 'Marunnu Pareekshichathu Ethical', *Malayala Manorama*. However, the report in *Deshabhimani* contradicted this. The report quoted Dr Krishnan Nair as saying that the RCC clinical trials had been approved by both the RCC ethical committee and the DCGI. See 'Nirodhicha Marunnukal RCCyil', *Deshabhimani*. The press note of the Health Ministry of the Government of India which was widely reported in the regional press later on said that the DCGI had approved the import of M4N from the JHU in February 2001. See 'RCCyile Nirodhitha Marunnu Prayogam Kendram Anweshikkunnu', *Malayala Manorama*, 31 July 2001, p. 9; 'Nirodhicha Marunnu Upayogichathu Anweshikkum', *Desha-bhimani*, 31 July 2001, p. 1. It was also mentioned that the approval was granted on the basis of the recommendations of the RCC ethical committee. See 'RCC Marunnu Pareekshanam Kendram Anweshikkunnu', *Kerala Kaumudi*, 31 July 2001, p. 1.

51. See Prasad (2009) for a detailed analysis of the politics of conducting clinical trials in developing countries like India.

52. 'RCCyile Nirodhitha Marunnu', *Malayala Manorama*; 'Nirodhicha Maru-nnu Upayogichathu', *Deshabhimani*; 'RCC Marunnu Pareekshanam', *Kerala Kaumudi*.

53. 'RCCyile Nirodhitha Marunnu', *Malayala Manorama*, p. 9; 'RCC Marunnu Pareekshanam', *Kerala Kaumudi*, 2001, p. 1.

54. Dr Purvish Parikh who was appointed as the single-member enquiry commission was Professor of medical oncology at the Tata Memorial Hospital, Mumbai, and was also the then secretary of the Indian Cooperative Oncology Network (ICON). 'Dr Parikh Anweshikkum', *Mathrubhumi*, 1 August 2001, p. 1; 'Cancer Centre Vivadam: Anweshikkan Utharavu', *Deshabhimani*, 1 August 2001, p. 10; 'Dr Parikh Anweshikkum', *Madhyamam*, 1 August 2001, p. 1; 'RCC Vivadam Anweshikkan Ekanga Commission', *Kerala Kaumudi*, 1 August 2001, p. 1.

55. 'Director Hajarakanam', *Mathrubhumi*, 1 August 2001, p. 1; 'RCC Directorkku Manushyavakasa Commissionte Summons', *Madhyamam*, 1 August 2001, p. 7; 'Dr Krishnan Nairkku Manushyavakasa Commissionte Summons', *Kerala Kaumudi*, 1 August 2001, p. 1.

56. 'Cancer Centre: Judicial Anweshanam Venam', *Mathrubhumi*, 1 August 2001, p. 4.

57. 'Parikhinte Anweshanavumayi Yojikkanavilla: Bhattathiri', *Madhyamam*, 1 August 2001, p. 7; 'RCC Vivadam Puthiya Vazhithirivilekku', *Deshabhimani*, 2 August 2001, p. 7.

58. 'RCC Vivadam Puthiya', *Deshabhimani*.

59. 'Cancer Centre: Anweshanam Nishpakshamakillennu', *Madhyamam*, 2 August 2001, p. 5.

60. 'Cancer Centre: Anweshanam', *Madhyamam*.

61. 'Arbuda Marunnu Pareekshikkan Anumathi Nalkiyirunnillennu Hopkins Sarvakalasala', *Mathrubhumi*, 1 August 2001, p. 1; 'RCC Vivadam Puthiya', *Deshabhimani*; 'Pareekshanathinu Anumathi Nalkiyillennu JHU: RCC Padanam Nirthiveykkan Gaveshakaykku Nirdesam', *Madhyamam*, 1 August 2001, p. 1.

62. 'Pareekshanathinu Anumathi Nalkiyillennu', *Madhyamam*.

63. 'Pareekshanathinu Anumathi Nalkiyillennu', *Madhyamam*.

64. Dr C.R. Soman (now late) was formerly Professor and the Head of the Department of Nutrition at the Government Medical College, Thiruvananthapuram. Later on, he became the director of Health Action by People, an action-research NGO in Thiruvananthapuram.

65. Dr M. Krishnan Nair, 'Americayil Nirodhichittilla', *Malayala Manorama*, 1 August 2001, p. 4.

66. Dr C.R. Soman, 'Niyama Lankhanam, Adharmikam', *Malayala Manorama*, 1 August 2001, p. 4. The discussion of Dr Soman's intervention is based on this article, if otherwise not mentioned.

67. The Forum for Patients' Rights is a civil society organisation which took a strong anti-clinical trials position in the controversy.

68. 'RCC Vivadam: Samagranweshanam Venamennu Charcha', *Madhyamam*, 2 August 2001, p. 7.

69. 'RCCyil Judicial Anweshanam Venam', *Mathrubhumi*, 2 August 2001, p. 3.

70. 'RCCyil Judicial Anweshanam', *Mathrubhumi*; 'RCC Vivadam: Samagranweshanam', *Madhyamam*.

71. 'RCC Vivadam: Samagranweshanam', *Madhyamam*.

72. 'Nirodhitha Marunnu Kuththi Vechathu Manushyavakasa Lankhanamennu', *Malayala Manorama*, 2 August 2001, p. 9.

73. 'RCC Vivadam: Samagranweshanam', *Madhyamam*.

74. 'RCC Vivadam: Samagranweshanam', *Madhyamam*.

75. 'RCC Vivadam: Samagranweshanam', *Madhyamam*; 'RCCyil Judicial Anweshanam', *Mathrubhumi*; 'Nirodhitha Marunnu Kuththi', *Malayala Manorama*.

76. See 'Marunnu Pareekshanam: Samvadathil Doctormarude Vakku Tharkkam', *Malayala Manorama*, 4 August 2001, p. 9; 'Marunnu Pareekshanam: Samvadathil Doctormar Cheri Thirinju Tharkkichu', *Mathrubhumi*, 4 August

2001, p. 2; 'Marunnu Pareekshanam: Samvada Velayil Tharkkavum Bahalavum', *Deshabhimani*, 4 August 2001; 'RCC Vivadam: Doctormar Thammil Tharkkam', *Madhyamam*, 4 August 2001, p. 7; 'RCC Prasnathil Doctors Cheri Thirinju', *Kerala Kaumudi*, 4 August 2001, p. 2.

77. 'RCC Vivadam: Doctormar', *Madhyamam*.

78. 'RCCkkethireyulla Pracharanathil Ninnu Pinmaranam: Mukhyamanthri', *Mathrubhumi*, 3 August 2001, p. 2; 'RCCye Nasippikkaruthu: Mukhyamanthri', *Kerala Kaumudi*, 3 August 2001, p. 3.

79. 'RCCye Nasippikkaruthu', *Kerala Kaumudi*.

80. RCCkkethireyulla Pracharanathil Ninnu', *Mathrubhumi*; 'RCCye Nasippikkaruthu', *Kerala Kaumudi*.

81. 'RCCye Nasippikkaruthu', *Kerala Kaumudi*.

82. 'RCCye Nasippikkaruthu', *Kerala Kaumudi*.

83. M.N. Vijayan, 'RCC', *Deshabhimani*, 33/10, 12 August 2001, p. 12. M.N.Vijayan was the chief editor of this leftist magazine owned by the CPI(M). In the next issue (33/11, 19 August 2001), the magazine published the RCC controversy as its cover story.

84. 'RCC: Mukhyamanthriyude Asahishnutha Janadhipathya Virudham: M.N. Vijayan', *Deshabhimani*, 5 August 2001, p. 4; 'RCC Vivadam: Mukhyamanthriyude Asahishnutha Janadhipathya Virudham: M.N. Vijayan', *Madhyamam*, 5 August 2001, p. 7.

85. 'RCC Vivadam: Mukhyamanthriyude', *Madhyamam*.

86. 'RCC: Mukhyamanthriyude Prasthavana Dourbhagyakaram', *Madhyamam*, 4 August 2001, p. 7.

87. 'RCCye Thakarkkan Sramam: Jeevanakkar', *Malayala Manorama*, 3 August 2001, p. 2; 'RCCye Thakarkkan Goodalochana', *Deshabhimani*, 3 August 2001, p. 8; 'RCC Directorku Pinthunayayi Doctorsum Staffum', *Kerala Kaumudi*, 3 August 2001, p. 11.

88. 'RCCkkethire Chilar Dooshitha Valayam Srishtikkunnu: Manthri Sankaran', *Malayala Manorama*, 4 August 2001, p. 9.

89. See 'RCCye Rakshikkan Nadapadi Venam: MLA', *Mathrubhumi*, 6 August 2001, p. 13; 'RCCye Thaaradichu Kattaruthu', *Mathrubhumi*, 7 August 2001, p. 11; 'Cancer Centreine Thakarkkaruthu', *Kerala Kaumudi*, 8 August 2001, p. 10; 'RCC Doctormar Aasayakkuzhappam Srishtiykkunnu: K.V. Surendranath', *Kerala Kaumudi*, 10 August 2001, p. 10; Vishnu Narayanan Nampoothiri, 'RCC Erinjudaykkaruthu', *Mathrubhumi*, 15 August 2001, p. 4, editorial page article; 'RCC Vivadam Sthapanathinte Nadathippinu Bheeshaniyakaruthu: ONV', *Malayala Manorama*, 17 August 2001, p. 17.

90. See 'RCC: Vivadam Venda', *Malayala Manorama*, 5 August 2001, p. 2.

91. 'RCC Vivadam: John Hopkins Anweshanam Thudangi', *Madhyamam*, 5 August 2001, p. 1.

92. 'RCC Vivadam: John Hopkins', *Madhyamam*.

93. '"Pareekshanamriga"makunnathu Thadayan Americayil Niyama Bhedagathi Varunnu', *Madhyamam*, 7 August 2001, p. 7; 'RCCyile Vivada Pareekshanam Angeekarichittillennu Hopkins', *Deshabhimani*, 7 August 2001, p. 10.

94. 'RCCyile Vivada Pareekshanam', *Deshabhimani*.

95. 'RCCyile Vivada Pareekshanam', *Deshabhimani*.

96. The journal article being quoted in the newspaper is possibly Bagla and Marshall (2001). The journal also took note of the action taken by the Ministry of Health and Family Welfare in another article. See Bagla (2001).

97. 'Marunnu Pareekshanathinu Hopkins Paththu Laksham Nalki', *Madhyamam*, 13 August 2001, p. 1.

98. 'Marunnu Pareekshanathinu Hopkins', *Madhyamam*.

99. 'Marunnu Pareekshanathinu Hopkins', *Madhyamam*.

100. 'Marunnu Pareekshanathinu Hopkins', *Madhyamam*. *Mathrubhumi* mentioned that the 'Quintiles International' was a CRO that operated from the United States. See 'Marunnu Pareekshanam: Kooduthal Videsa Agencykalkku Panku', *Mathrubhumi*, 12 August 2001, p. 5. For similar allegations regarding the involvement of a Singapore-based drug company (seemingly BioCure Medicals), see 'Marunnu Pareekshanam: Kooduthal Samsayangal', *Mathrubhumi*, 3 August 2001, p. 5.

101. 'RCC Vivadam IMA Sangham Anweshikkum', *Mathrubhumi*, 3 August 2001, p. 2; 'RCC: IMA Theliveduppu Thudangi', *Deshabhimani*, 5 August 2001, p. 4; 'IMA Commission Theliveduppu Thudangi', *Malayala Manorama*, 5 August 2001, p. 9.

102. The Prathikarana Vedi ('Response Forum') was launched at Thiruvananthapuram with an intention to fight corruption. Reputed media analyst B.R.P. Bhaskar was the president of the forum. See 'Azhimathikkethire Prathikarana Vedi Roopeekarichu', *Kerala Kaumudi*, 4 July 2001, p. 10.

103. 'RCC Vivadam IMA Sangham Anweshikkum', *Mathrubhumi*, 3 August 2001, p. 2; 'RCC Vivadam: Sarkar Committee Aparyapthamennu Prathikarana Vedi', *Madhyamam*, 5 August 2001, p. 3.

104. 'RCC, Karsana Nadapadi Sweekarikkkanam: Human Rights Protection Council', *Madhyamam*, 9 August 2001, p. 9; 'RCC: Vazhivitta Pareekshanaththinethire Nadapadi Venam: Sasthra Parishad', *Kerala Kaumudi*, 8 August 2001, p. 2; 'RCC Pareekshanam Adharmmikamennu', *Kerala Kaumudi*, 12 August 2001, p. 2; 'Uttharavadikale Matti Nirthi Anweshikkanam: Parishad', *Deshabhimani*, 8 August 2001, p. 6; 'RCC Pareekshanam: Nadapadi Venamennu Parishad', *Malayala Manorama*, 8 August 2001, p. 9.

105. 'Clinical Pareekshana Niyamangal Cancer Centre Avaganichu: Padana Report', *Madhyamam*, 7 August 2001, p. 3.

106. 'Clinical Pareekshana Niyamangal', *Madhyamam*.

107. Seminars organized by the Mukthi, the IPA-Kerala, the KSSP, and the Kerala M. Pharm Students' Forum were widely reported in the regional press. See news reports on 4 August 2001 and 8 August 2001.

108. The KSSP was the most vocal speaker in this regard. It also strongly demanded that the ICMR guidelines should be made applicable for all medical science research institutes in the country. See RCC Pareekshanam: Nadapadi', *Malayala Manorama*; 'Uttharavadikale Matti Nirthi', *Deshabhimani*; 'RCC: Vazhivitta Pareekshanaththinethire', *Kerala Kaumudi*.

109. See Varughese (2014) for a detailed analysis of alternative models of technological governance proposed by social movements in India.

110. He was under Dr Gangadharan's treatment. '"Guneappanni" Ayathu Aarkku Vendiyennariyathe Gopal', *Madhyamam*, 3 August 2001, p. 5.

111. 'Gopal Manushyavakasa Commissionu Parathi Nalki', *Madhyamam*, 3 August 2001, p. 2.

112. 'Gopalil Pareekshanam Nadathiyathu Chattangal Lankhichu', *Madhyamam*, 18 August 2001, p. 2.

113. 'Nirodhitha Marunnu Kuththi Vecha Randu Per Neraththe Marichennu Mozhi', *Madhyamam*, 11 August 2001, p. 2; 'Marunnu Pareekshanam: Randu Per Marichathayi Aaropanam', *Deshabhimani*, 11 August 2001, p. 1.

114. 'Marunnu Pareekshanathil Amma Marichathayi Makante Parathi', *Deshabhimani*, 12 August 2001, p. 1; 'Amma Marichathu Marunnu Pareekshanam Moolamenna Aropanavumayi Makan', *Madhyamam*, 12 August 2001, p. 3.

115. 'Rogi Marichathu Marunnu Pareekshanam Kondalla: RCC', *Mathrubhumi*, 14 August 2001, p. 7.

116. 'RCCye Thakarkkan Sramam', *Malayala Manorama*.

117. 'RCCye Thakarkkan Sramam', *Malayala Manorama*.

118. As pointed out by a social activist (name withheld) the author interviewed on 26 April 2006, the patients who underwent the clinical trials had been from poor socio-economic background, and hence were availing financial assistance through the welfare schemes run by the RCC. As chronic patients, they were also completely at the mercy of their doctors. These factors appear to have helped the RCC authorities to prevent the patients' voices from entering the scientific public sphere.

119. See 'RCCyude Visvasyatha Nashtappedutharuthu', *Mathrubhumi*, 19 August 2001, p. 3; See also the personal requests made by some of the signatories; Vishnu Narayanan Nampoothiri, 'RCC Erinjudaykkaruthu', *Mathrubhumi*, 15 August 2001, p. 4, editorial page article; 'RCC Vivadam Sthapanathinte', *Malayala Manorama*, p. 17.

120. For a detailed discussion, see Martin (1988). O.N.V. Kurup, 'RCCye Nasippikkaruthe!', *Mathrubhumi*, 17 August 2001, p. 4, letter to the editor;

Sugathakumari, Thiruvananthapuram, 'RCCyude Agnipareeksha', *Mathrubhumi*, 9 August 2001, p. 4, letter to the editor.

121. See 'RCCkkethire Pracharanam Nadathiyavare Pirichu Vidanam', *Deshabhimani*, 20 August 2001, p. 8.

122. See, for example, Dr K. Balaraman, 'RCC: Kolahalam Apalapaneeyam', *Kerala Kaumudi*, 7 August 2001, p. 4, letter to the editor; Dr N.M. Muhammad Ali, 'Verittoru Sabdam', *Kerala Kaumudi*, 19 August 2001, p. 4, letter to the editor; Dr T.P. Gopalakrishnan, Kawadiyar, Thiruvananthapuram, 'Doctormar Porvili Nirthanam', *Mathrubhumi*, 23 August 2001, p. 4, letter to the editor; Ramesh G., Kozhikode, 'RCCyude Salpperu Nilanirthan', *Mathrubhumi*, 14 September 2001, p. 4, letter to the editor.

123. For a detailed discussion, see Martin (1988).

124. 'Gaveshanam Kurisil Thoongunnu', *Kerala Kaumudi*, 14 August 2001, p. 4.

125. 'Gaveshanam Kurisil Thoongunnu', *Kerala Kaumudi*.

126. *Mathrubhumi* also used its 'Letters to the Editor' column in a similar fashion, but more benignly. See, for example, Sugathakumari, Thiruvananthapuram, 'RCCyude Agnipareeksha', *Mathrubhumi*, 9 August 2001, p. 4, letter to the editor; O.N.V. Kurup, 'RCCye Nasippikkaruthe!', *Mathrubhumi*, 17 August 2001, p. 4, letter to the editor. The construction of the 'Letters to the Editor' columns by the press to further its interests is a well-noticed fact. See Richardson and Franklin (2003). For a detailed discussion on the letters to the editor as a genre, see Chapter 7.

127. Adv. V. Sadasivan, 'Cancer Centre Vivadam', *Kerala Kaumudi*, 5 August 2001, p. 4, letter to the editor; Dr K. Balaraman (former director, Dept. of Health), Thiruvananthapuram, 'RCC: Kolahalam Apalapaneeyam', *Kerala Kaumudi*, 7 August 2001, p. 4, letter to the editor.

128. See, for example, Umayannur E. Asseem Kunju, Perayam, 'RCCye Nasippikkaruthu', *Kerala Kaumudi*, 6 August 2001, p. 4, letter to the editor.

129. Dr N.M. Muhammad Ali, Thiruvananthapuram, 'Verittoru Sabdam', *Kerala Kaumudi*, 19 August 2001, p. 4, letter to the editor; Dr L. Lalitha Kumari, 'RCCkku Ethire Gooda Padhathi', *Kerala Kaumudi*, 3 September 2001, p. 4, letter to the editor. *Mathrubhumi* published a letter that criticized the free-for-all contest in full public view by the RCC scientists. See Dr T.D. Gopalakrishnan, Kawadiar, 'Doctormar Porvili Nirthanam', *Mathrubhumi*, 23 August 2001, p. 4, letter to the editor.

130. Priyanjana Prabhakar, Thiruvananthapuram, 'Athu Mathramo Nirodhikkappetta Oushadham?', *Kerala Kaumudi*, 5 August 2001, p. 4, letter to the editor.

131. Dr M.P. Mani, Kudamaloor, 'Gaveshanangal Sarvakalasalakalil Mathramakanam', *Kerala Kaumudi*, 21 August 2001, p. 4, letter to the editor.

132. Dr C.R. Soman, 'Cancer Centreine Apakeerthippeduthukayalla Lakshyam', *Kerala Kaumudi*, 12 September 2001, p. 4, letter to the editor.

133. Dr P.A. Kurup, Thiruvananthapuram, 'Marunu Pareekshanam: Reportinu Kakkuka', *Mathrubhumi*, 23 August 2001, p. 4, letter to the editor.

134. The commission was constituted by the Union Ministry of Health and Family Welfare. The commission consisted of Dr A.R. Singh (representative of the DCGI), Dr Vasantha Muthu Swami (ICMR), and Dr A.N. Serin (Consultant, Department of Paediatrics, Nirman Bhavan). See 'Cancer Centre Vivadam: Kendra Sangham Ethi', *Madhyamam*, 9 August 2001, p. 5.

135. For example, the commission appointed by the union Health Ministry was reported to have not had a proper orientation regarding the mode of enquiry. The neutrality of the commission also was seriously doubted. See 'Cancer Centre Vivadam', *Madhyamam*. The *Mathrubhumi* also reported the ambiguity regarding the aims and objectives of the enquiry. 'RCC Anweshana Sangham Ethi', *Mathrubhumi*, 9 August 2001, p. 5.

136. 'Anweshana Commissionu Sancharikkan RCC Vahanam', *Madhyamam*, 10 August 2001, p. 2.

137. 'Anweshana Commissionu Sancharikkan', *Madhyamam*. See also 'RCC-yude Vahanam Sarkar Nalkiyathennu Thettidharichu', *Madhyamam*, 11 August 2001, p. 7.

138. 'Anweshana Commissionu Sancharikkan', *Madhyamam*.

139. 'RCC Anweshanam Nale Thudangum', *Deshabhimani*, 8 August 2001, p. 6.

140. 'Vidagdha Samithi Anweshikkanam', *Deshabhimani*, 8 August 2001, p. 6; 'Utharavadikale Matti Nirthi', *Deshabhimani*. The Youth Congress Medical College Committee activists also staged an aggressive demonstration against the commission on 10 August in front of the Medical College auditorium where the hearing was being held. 'RCC: Commissionethire Youthkar', *Deshabhimani*, 11 August 2001, p. 10; 'RCCyude Vahanam Sarkar', *Madhyamam*.

141. 'Vivada Marunnu Pareekshikkappetta Rogikalil Ninnu Thelivedukkum', *Kerala Kaumudi*, 11 August 2001, p. 3.

142. 'RCC Vivadathinu Pinnil Gooda Lakshyam: Aropanavumayi Vannavar Swantham Asupathrikku Thudakkamittavar', *Kerala Kaumudi*, 22 August 2001, p. 1.

143. The news report also mentioned that some more persons from the RCC were involved in the venture.

144. 'RCC-Paraathiyil Puthiya Society Rekhakal Mudra Vechu', *Malayala Manorama*, 22 August 2001, p. 13. The report explained that seven doctors from the RCC were involved in the registration of the society.

145. Only two newspapers reported it on the next day, but more joined in over the next few days. 'Cancer Society Rekhakal Mudra Vechathu Krithrimam Ozhivakkan', *Mathrubhumi*, 23 August 2001, p. 18; 'RCC Vivadam: Swantham Aasupathri Thudangan Lakshyamittathum Anweshikkum', *Deshabhimani*, 23 August 2001, p. 3.

146. 'RCCkkethire Pracharanam', *Deshabhimani*; 'RCC: Nalu Perkkethire Parathi', *Deshabhimani*, 24 August 2001, p. 3; 'RCC Protection Forum Prakshobhathilekku', *Kerala Kaumudi*, 28 August 2001, p. 3; The forum went a step ahead when it filed a petition against Dr C.R. Soman who had spoken against the clinical trials, alleging that he had indulged in a slander campaign against the RCC. See 'RCC Vivadam: Dr. C.R. Somanethire Protection Forum Paraathi Nalki', *Deshabhimani*, 9 September 2001, p. 3.

147. 'Cancer Centreinu Hopkins Nalkiyathu Naalpathu Lakhsham: Thuka Prathyeka Accountil', *Madhyamam*, 28 August 2001, p. 1. The news report published a photograph of the receipt of the money sent by the manager of the bank to the RCC director as evidence for the same.

148. Dr Bhattathiri and Dr Gangadharan resigned from the RCC following the closure of the controversy.

149. 'RCC: Marunnu Pareekshanam Nirthan Kendra Nirdesam', *Mathrubhumi*, 10 September 2001, p. 1, lead article; 'RCC: Doctormarkkethire Nadapadikku Sadhyatha', *Malayala Manorama*, 10 September 2001, p. 1; 'RCCyile Pareekshanangal Aru Masathekku Suspend Cheythu', *Deshabhimani*, 10 September 2001, p. 1; 'RCCyil Gaveshanam Nirthivekkan Kendra Nirdesam', *Madhyamam*, 10 September 2001, p. 1; 'RCCyodu Kendram Visadeekaranam Thedum', *Kerala Kaumudi*, 16 September 2001, p. 12.

150. 'RCC: Marunnu Pareekshanam', *Mathrubhumi*; 'RCCyil Gaveshanam Nirthivekkan', *Madhyamam*; 'RCCyile Pareekshanangal Aru', *Deshabhimani*.

151. 'RCCyile Pareekshanangal Aru', *Deshabhimani*.

152. 'RCCyile Pareekshanangal Aru', *Deshabhimani*; 'RCC: Doctormarkkethire Nadapadikku', *Malayala Manorama*.

153. 'RCC: Doctormarkkethire Nadapadikku', *Malayala Manorama*; 'Anumathiyillathe Vivada Marunnu Konduvannathu Pradhana Pizhavu', *Malayala Manorama*, 10 September 2001, p. 11.

154. 'Anumathiyillathe Vivada Marunnu', *Malayala Manorama*.

155. 'RCC: Marunnu Pareekshanam', *Mathrubhumi*; 'RCCyil Gaveshanam', *Madhyamam*; 'RCCyile Pareekshanangal Aru', *Deshabhimani*.

156. 'RCCyile Pareekshanangal Aru', *Deshabhimani*. Earlier the 12 guidelines of the ICMR were applicable only to the research institutions affiliated to the ICMR.

157. 'RCC: Doctormarkkethire Nadapadikku', *Malayala Manorama*.

158. 'Nirodhitha Marunnu Upayogichittilla: Chattalankhanathinu Nadapadi, Doctormarkkethire Nadapadiyilla', *Mathrubhumi*, 16 September 2001, p. 5; 'RCC Doctorsinu Ethire Nadapadi Undavilla', *Madhyamam*, 16 September 2001, p. 5; 'RCCyodu Kendram Visadeekaranam', *Kerala Kaumudi*.

159. 'Marunnu Pareekshanam: Nadapadi Paalichillennu IMA Report', *Malayala Manorama*, 14 September 2001, p. 9; 'Cancer Centreile Marunnu

Pareekshanam Krama Virudhamennu IMA Report', *Madhyamam*, 14 September 2001, p. 9; 'Marunnu Pareekshanam: Nadapadikramam Paalichilla', *Mathrubhumi*, 15 September 2001, p. 11; 'IMA Report: RCCye Rakshikkan Sramam', *Madhyamam*, 15 September 2001, p. 7; 'RCC Nadapadikramam Paalichilla: IMA', *Kerala Kaumudi*, 15 September 2001, p. 12.

160. 'RCC Prasnam Othukki Theerkkunnu', *Deshabhimani*, 16 September 2001, p. 7; 'IMA Report: RCCye Rakshikkan Sramam', *Madhyamam*, 15 September 2001, p. 7.

161. 'Manushyavakasa Lankhanam Illenna Vadam Mandatharam', *Mathrubhumi*, 16 September 2001, p. 5; 'Marunnu Pareekshanam Manushyavakasa Lankhanam: Dr. Bhattathiri', *Madhyamam*, 16 September 2001, p. 5.

162. *Malayala Manorama* was the only newspaper which reported the content of the preliminary report. Others published the news when the final report was submitted to the government one month later.

163. 'Nirodhitha Marunnu Cancer Centreil Pareekshichittillennu Anweshana Report', *Malayala Manorama*, 5 October 2001, p. 9; 'RCCyil Nirodhitha Marunnu Pareekshichillennu Anweshana Commission', *Madhyamam*, 18 November 2001, p. 6.

164. 'Nirodhitha Marunnu Cancer', *Malayala Manorama*.

165. 'Nirodhitha Marunnu Cancer', *Malayala Manorama*; 'Vivada Marunnu Nirodhichathallennu Report', *Mathrubhumi*, 18 November 2001, p. 1.

166. 'Nirodhitha Marunnu Cancer', *Malayala Manorama*.

167. 'Nirodhitha Marunnu Cancer', *Malayala Manorama*; 'RCCyil Nirodhitha Marunnu', *Madhyamam*.

168. 'RCC: Mukhya Manthri Vittu Nilkkanam', *Deshabhimani*, 16 November 2001, p. 6; 'RCCye Doctormarude Lobbykku Vittu Kodukkaruthu', *Deshabhimani*, 6 December 2001, p. 12; 'RCC Marunnu Pareekshanathinu Pinnil Pravarthichavarkkethire Nadapadi Venam', *Madhyamam*, 24 November 2001, p. 3.

169. 'Nirodhitha Marunnu Cancer', *Malayala Manorama*; 'Marunnu Pareekshanam: Prof. Ru Chihye Baliyaadakkan Aniyara Neekkam', *Madhyamam*, 19 November 2001, p. 5.

170. See Dr Raveendran Kummil, 'RCC Vivadavum Vasthuthakalum', *Kerala Kaumudi*, 30 November 2001, p. 4, editorial page article; 'RCC Vivadam Arthasoonyam: Dr. Padmanabhan Nair', *Kerala Kaumudi*, 11 February 2002, p. 2. The latter news report was based on an interview with Dr Padmanabhan Nair, who was the former director of the Department of Biomedical Research at JHU. He claimed that it was he who introduced Dr Ru Chih Huang to Dr Krishnan Nair.

171. The conclusions of the JHU Commission were reported in the Malayalam press. 'Marunnu Pareekshanam Hopkins Sammathichu: Sastrajjnaykku

Vilakku', *Deshabhimani*, 14 November 2001, p. 1; 'RCC Marunnu Pareekshanam Anumathiyillathe: Hopkins Samithi, Gaveshakaykku Vilakku', *Madhyamam*, 14 November 2001, p. 1.

172. The state secretary of the DYFI, T. Sasidharan, pointed out that the JHU enquiry report made it clear that the clinical trials were unethical and he criticized the lack of will power on the part of the state government to take action against the harmful rackets witnessed at the RCC. 'RCCye Doctormarude Lobbykku Vittu Kodukkaruthu', *Deshabhimani*, 6 December 2001, p. 12.

173. 'Marunnu Pareekshanam: Nadapadikramam Paalichittillennau Manthri Thakur', *Malayala Manorama*, 4 December 2001, p. 9; 'Vivada Marunnu Pareekshanathil Manushyavakasa Lankhanamilla: Kendra Manthri', *Mathrubhumi*, 4 December 2001, p. 5; 'RCCyil Nirodhicha Marunnu Pareekshichittilla: Manthri', *Deshabhimani*, 4 December 2001, p. 5.

174. 'RCCyil Nirodhitha Marunnu Upayogichittilla: Manthri', *Malayala Manorama*, 4 December 2001, p. 9; 'RCCyil Nirodhitha Marunnu Upayogichittilla: Manthri', *Mathrubhumi*, 4 December 2001, p. 7; 'Nirodhitha Marunnu Chikilsakku Upayogichittillennu Manthri', *Deshabhimani*, 4 December 2001, p. 8; 'RCCye Sikshikkan Nivruthiyilla: Manthri', *Kerala Kaumudi*, 30 November 2001, p. 4; 'Prathipaksham Avakasa Lankhanathinu Noticinu Orungunnu', *Madhyamam*, 30 November 2001, p. 5; 'RCC Prasnam: Dr. Parikhinte Report Thallanam', *Deshabhimani*, 5 December 2001, p. 9.

175. 'RCC Prasnam: Dr. Parikhinte', *Deshabhimani*.

176. 'RCC: Rekhaa Moolam Paraathi Labhichal Anweshikkumennu', *Deshabhimani*, 8 December 2001, p. 8. The Petitions Committee of the Legislative Assembly was appointed by the Speaker of the House to investigate the complaints against the RCC. See 'Sabha Samithi Anweshikkum', *Malayala Manorama*, 19 December 2001, p. 9; 'RCCkkethiraya Parathi Petitions Committee Anweshikkum', *Deshabhimani*, 19 December 2001, p. 1.

177. 'RCCye Kutta Vimukthamakkunna Report Manthri Sabha Angeekarichu', *Malayala Manorama*, 14 February 2002, p. 9; 'RCC Anweshana Report Angeekarichu', *Deshabhimani*, 14 February 2002, p. 2.

178. Quintiles India, the company that was involved in the RCC clinical trials, clearly states why India is a target for them to carry out clinical trials: 'India has a vast population. Patient access is fast; fully a third of the population lives in urban areas. Large portions of the population have not been exposed to prior treatment. Tropical diseases and diseases of developed countries are both common in India. Trials opportunities include cardio-vascular diseases, diabetes, degenerative neurological diseases, cancers, psychiatric illnesses, and infectious diseases.' Quoted in Prasad (2009: 5) from www.quintiles.com/Corporate_Info/Regions/south_africa_and_india/India/India.htm, which the author accessed in April 2007 and is not available now.

179. The KSSP's interventions in the controversy, though limited, were contingent upon its long history of engagement with issues of public health and medical research. Dr Joy Elamon who represented the KSSP in the RCC controversy also corroborated this in a personal interview on 24 April 2006.

5 Loss of Trust in Experts
Earthquakes, Dams, and Well Collapses

U nlike the scientific controversy regarding the clinical trials in the Regional Cancer Centre (RCC) in Thiruvananthapuram that had originated within the scientific community and then erupted into the public domain, the present chapter delineates a controversy that had its origin in the public domain itself. The main reason behind the emergence of the controversy was the stark contradictions and ambivalences in the expert advice provided by scientists belonging to different institutional backgrounds following public disquiet over a series of earthquakes and allied geophysical phenomena in Kerala in 2001. While the scientists themselves contacted the media to push their agendas in the previous case, here it was the scientific-citizen public who demanded the intervention of scientists. Unlike the RCC controversy that debated the regulation of drug research and public governance of science, here the deliberations were steered by a gradual erosion of public trust on scientific interpretations. The present case, hence, gives us a rare opportunity to understand the conflicts between experts and the scientific-citizen public in perceiving the risks of micro-earthquakes and allied phenomena in the region, and the active role of media in initiating and sustaining public deliberations.

The scientific controversy under discussion was kick-started when an earthquake of low intensity occurred in Kerala on 12 December 2000. Many tremors of lower intensity continued striking the region from then on, and this was followed by the emergence of a conflict between the experts and the public regarding the interpretation of the events.

The controversy intensified following a massive earthquake in Gujarat on 26 January 2001 when reports appeared in the press linking it with the tremors in Kerala, contrary to the assurances of the experts. The fear of a looming disaster and the tremors as its precursors gained further momentum when a range of unnatural incidents occurred, among which well collapses were the most prominent, leading to more conflicts among various teams of researchers as well as between experts and the public. The chapter will focus on the dynamic role of the media and the professional relationship between scientists and journalists in shaping the risk discourse in the context of these events in Kerala.

The earthquake on 12 December 2000 struck nine districts of Kerala with medium intensity, a first one at around 6.54 a.m. and a second one at 8.20 a.m., as reported by newspapers the next day.[1] All the five newspapers under analysis carried the news as their headlines, indicating the general public perception of earthquakes/micro-tremors as posing great risk in everyday life.[2] The newspapers provided a wide range of scientific perspectives and information on the tremors from the very first day. Their reliance on various sources of scientific information and the portrayal of these sources as being opposed to each other were central to the creation of the controversy.

The debates of the day mainly revolved around three issues. The first was a polemic regarding the magnitude and epicentre of the quake. Scientists from the Centre for Earth Science Studies (CESS),[3] Thiruvananthapuram, argued that the tremor was of an intensity of 5 on the Richter scale, and suggested Painavu, the district headquarters of Idukki, as the epicentre. On the other hand, the Kerala State Electricity Board (KSEB)'s geological research wing argued that the tremor was of a lower magnitude of 3.35, and according to them the epicentre was Melukavu, a hill station in Kottayam district. A second issue that cropped up was the idea that reservoirs induced earthquakes. Most of the newspapers presented this argument citing scientific papers and quoting different scientists. The issue acquired importance because Idukki is the most 'dammed' district in Kerala[4] and, as suggested by the CESS scientists, the epicentre was Painavu, where Idukki arch dam—a part of one of the major hydroelectric power projects in the state—is located. A third issue related to the controversy was the impact of the micro-earthquakes on the water reservoirs in the region. The foci of the dispute were the Mullaperiyar dam, which was built during the colonial period, and the Idukki arch dam, which is considered as a modern engineering marvel. These three moot points were the axes of the public controversy that raged through the following days.

Dispute over the Epicentre and Intensity of the Earthquake

When the disagreements in deciding the epicentre and magnitude of the earthquake appeared in the press, the scientific public sphere became dynamic with the involvement of a wide array of actors. Newspapers turned into active participants in the deliberation, criticizing the scientific community for their lack of consensus over the technical interpretation of the natural phenomena. Newspapers like *Mathrubhumi* strongly critiqued the contradictions in the findings of the CESS and the KSEB in the context of a second set of low-intensity tremors that shook the region on the very next day, that is, 13 December 2000.[5] The common fear that was reflected in the newspaper reports was the possibility of these micro-earthquakes being indicators of an impending disaster. The regional press highlighted the ambiguities in the CESS scientists' argument that no major earthquake would occur in Kerala. *Madhyamam* contended that Kerala had become prone to earthquakes, contrasting the findings from earlier investigations of the Department of Mining and Geology (DMG) with those of the CESS.[6]

Amidst escalating public ire on the contradictions of scientific interpretations, the CESS and the geological research wing of the KSEB, the two adversaries in the controversy, sensed the dangers involved and started defending their respective claims in the media using different strategies. The intervention of the regional press thus literally dragged the experts into defending themselves in public. The CESS scientists began investigations on 14 December in Painavu in order to determine the epicentre with greater precision.[7] The KSEB, on the other hand, assisted the expert team from the India Meteorological Department (IMD). Although the expert team from IMD[8] began their investigation at Melukavu, the epicentre according to the KSEB, it is quite interesting to see that they never made their association with the KSEB explicit, and presented themselves in a non-partisan manner and upheld scientific objectivity.[9] They argued that their objective was to decide the epicentre accurately by taking into consideration both the technical interpretations.[10] In a display of impartiality and high scientific credentials, they visited Painavu as well. They concluded in their technical statement that the epicentre could be in between Melukavu and Painavu—perhaps as a national institute, they saw themselves playing the role of an adjudicating authority with higher epistemological power.

The intervention of the IMD as a national scientific institution with greater legitimacy failed to end the controversy, thanks to the media's contestation of the IMD's claim to neutrality as an adjudicator.[11] *Mathrubhumi* attempted to play up the dispute at this juncture by alleging that the KSEB's findings were based on data obtained from only 4 out of its 12 seismic stations.[12] This happened, the newspaper claimed by quoting KSEB officials, because of a national strike observed by the KSEB employees on 12 December that made the data from other stations unavailable. The newspaper was apparently trying to cash in on the mounting public disapproval of political strikes and *hartals*,[13] another debate that intermittently appears in the regional press. However, this attempt was foiled by the left-oriented KSEB Workers' Association (CITU) with the support of *Deshabhimani*,[14] pointing out that in spite of the strike, the engineers had collected the accelerograph readings from all the centres and had accurately calculated the magnitude and the epicentre.[15] It criticized the ulterior motives of *Mathrubhumi*.[16] This has also suggested the framing of the dispute over magnitude and epicentre as one between the engineers of KSEB and the earth scientists of the CESS.

The tremors continued to occur in the region after the first wave of earthquakes on 12 and 13 December, and so did the dispute. The technical confrontation between the CESS and the KSEB intensified when the spokespersons of the institutions appeared with more evidence in support of their particular interpretations and, to strengthen their arguments, sought the help of other scientific institutions. The Meteorological Centre at Thiruvananthapuram supported the CESS.[17] The Geological Survey of India (GSI) entered the controversy by sending a research team for further investigations.[18] The DMG submitted its scientific report to the state government stating that the intensity was 3.5, supporting the KSEB's claim.[19] Meanwhile, the National Geophysical Research Institute (NGRI), Hyderabad, stated that according to calculations based on their recordings the intensity was 4.9, an explanation more in conformity with the findings of the CESS.[20] The research coalitions that the regional institutions had formed with the more 'powerful' national scientific institutions helped them to arrogate more epistemological legitimacy to their respective claims. However, this never yielded a resolution to the dispute itself, as we shall see later in the chapter.

Reservoir-Induced Seismicity: Idukki Dam as a Potential Threat

Another dispute running parallel to and tightly woven with the magnitude–epicentre controversy was the one over the argument that reservoirs could trigger earthquakes. From the very first day, serious questions concerning the role of the large number of reservoirs in Idukki district in inducing seismic activity in the region had been raised in the regional press. This concern acquired much importance thenceforth, when the newspapers began publishing more journalistic reports on the subject, drawing on different sources of expertise. As mentioned earlier, Idukki district has several dams located within a very small geographical area and this became a cause for public alarm.[21] The argument of the CESS that the epicentre was at Painavu, the location of the huge reservoir of the Idukki Hydroelectric Power Project (hereafter, Idukki dam), further reinforced the fear that the earthquake in the region had been induced by reservoirs.[22] *Malayala Manorama* suggested in this context the collapse of 'the long-standing belief of scientists' that Kerala was located in the safer 'South Indian Shield' and was, therefore, less prone to major earthquakes.[23] The newspaper discussed the pros and cons of the reservoir-induced seismicity (RIS) argument. *Kerala Kaumudi* supported the RIS hypothesis while reporting the small tremors that continued to occur at the Idukki dam site.[24] *Madhyamam* quoted scientists from the CESS to hint that the Idukki dam could be the immediate trigger for the earthquake.[25] The newspaper published three major analytical reports which examined the role of the Idukki dam in inducing earthquakes.[26] It took a pro-RIS stance and quoted existing studies and interviewed reputed scientists to drive home the point. A close perusal of the news reports reveals that the regional press in general shared this belief and suggested the possible threat of the dam bursting due to increased seismic activity in the area.

The argument of the scientists from the CESS regarding the epicentre and magnitude of the earthquake might have contributed to the commencement of the RIS controversy; and they were reported as endorsing the RIS hypothesis in the interviews given to some journalists.[27] The claim regarding the role of the Idukki dam in rendering the region earthquake prone was very much in the air since day one, and questions regarding the potential threat to Idukki dam in the likelihood of another earthquake were raised.[28] The CESS scientists who primarily suggested that the vicinity of the Idukki dam as the epicentre, thus, faced an unanticipated predicamaent that might have far-reaching consequences for them as employees in a governmental institution. In order to resolve this, they had to emphasize that the dam would not be affected by earthquakes, pointing out the modern engineering of the dam and the tectonic study that had been undertaken before its construction.[29] They mounted their argument carefully with the intention of shifting the attention to another reservoir at Mullaperiyar, which had been an eyesore for Kerala since the 1970s.[30] The scientists pointed out that the earthquakes must have damaged Mullaperiyar, the oldest dam in Kerala, built using 'pre-modern technology'.[31] Their strategy was to highlight Mullaperiyar dam, a late-nineteenth-century, colonial engineering marvel, as 'pre-modern' and (hence) vulnerable, by contrasting it with Idukki dam as a 'modern' technological artefact that can withstand earthquakes due to the advanced engineering practices involved in its construction in the 1970s. While the CESS scientists were trying hard to transfer the RIS debate to the context of the Mullaperiyar dam in order to divert public attention from the Idukki dam, the same objective, interestingly, had already been accomplished by the KSEB scientists who located the epicentre at Melukavu, a place quite far from the Idukki dam. The argument that the magnitude of the earthquake was lower further served their purpose. As scientists in the KSEB responsible for answering any questions regarding the safety of Idukki dam, they possibly might have anticipated the problem much ahead of the CESS scientists, and therefore vigorously defended their 'oppositional' findings.[32]

Even though the Mullaperiyar issue was the focus of deliberations, the public fear about the Idukki dam was equally strong. In its editorial on 14 December, *Madhyamam* challenged the assumptions of the CESS scientists for trivializing the possible threat of the Idukki dam.[33] The editorial pointed out that the statements of the experts were not based on thorough research. The newspaper also suggested that the scientists could not be relied upon due to inconsistencies in their scientific explanations, pointing out the difference of opinion that existed among them regarding the epicentre and the magnitude of the earthquake. The editorial cited studies supporting the RIS hypothesis and pointed out the cracks that developed in the Koyna dam site following the earthquake of 1967 as solid evidence.[34] The newspaper critiqued the anti-people and anti-environmental developmentalism in India that supported building big reservoirs even in ecologically and seismically sensitive areas like the Himalayan region.[35] *Mathrubhumi* too took a similar position, arguing that big dams were products of the unholy alliance between politicians, bureaucrats, and contractors.[36] It also pointed out that there were no scientific studies available on RIS in Kerala, and drew attention to the risk posed by the Idukki dam, suggesting that since 1974 there had been many dam-induced tremors near its catchment area.[37] *Mathrubhumi* refuted the argument of experts that reservoir-induced tremors would come to an end a few years after the dam's construction, and insisted on rethinking the develpomentalist ideas on which the construction of big dams had been justified and legitimized in the country. The newspaper also contended that scientists in Kerala were forced to be silent on the issue of RIS, pointing a finger to the evil nexus between politicians, bureaucrats, and contractors that impeded further scientific research on RIS in Kerala. *Mathrubhumi* seriously pushed their argument on RIS over the next few days and invoked the wider debate on ecological risks raised by large dams in the country.[38] This emphasis on the ecological risk posed by big dams was presented by linking the issue of tremors with the ecological and anti-developmentalist concerns that crystallized through a long deliberative process commencing in the late 1970s and 1980s in Kerala.[39]

The controversies regarding the epicentre and magnitude of the earthquake as well as the debate on whether reservoirs induce seismic activity reached a closure when Dr B.K. Rastogi, the assistant director of the NGRI, visited the earthquake-affected districts.[40] As a renowned scientist from a scientific institution of national repute, his intervention had greater public acceptance and he was considered as a supra-scientific authority capable of settling the dispute. He opined that the occurrence of a big earthquake in Kerala was unlikely although micro-seismic activity might continue for a couple of months.[41] He asserted that the earthquake did not affect the Idukki dam, and said that he was not sure about its impact on the Mullaperiyar dam.[42] He pointed out that the Idukki dam was only 24 years old and strong enough, but the latter was more than a century old and therefore weak.[43] Rastogi also refuted the RIS argument by pointing to the absence of scientific research that establishes a link between reservoirs and earthquakes.[44] Rastogi made an effort to vindicate the Idukki dam by carefully shifting the site of the controversy to the Mullaperiyar dam.[45] The research team of the GSI also substantiated Rastogi's exoneration of the Idukki dam later in their report to the state government.[46]

Rastogi intervened in the epicentre–magnitude controversy as well. He declared that the intensity of the earthquake was 5 on the Richter scale, and suggested that the discrepancy in the readings occurred because the CESS and the KSEB had used different formulae to calculate the magnitude.[47] According to him, the KSEB was using the old, out-dated formula and, therefore, the new formula should be adopted after conducting further research in the area so that more accurate calcula-tions would be possible.[48] He also suggested that more sophisticated instruments must be installed at the KSEB research stations in order to get more accurate readings.[49]

Though Rastogi's statements may appear opposed to the claims of the KSEB in a quick reading, it was not actually so. He was in close interaction with the KSEB research team during the visit.[50] It was clear that his conclusions were based on the data obtained from the KSEB's seismographs in Idukki.[51] Further, instead of supporting the argument of the CESS scientists, he was taking a nuanced position on the contro-versy. The issue of magnitude was resolved without criticizing the KSEB researchers but pointing out to the use of different formulas by the two research teams, and the lack of the latest scientific instrumentation at the KSEB research centre that were needed to make accurate recordings. Thus, by pointing out to the lack of infrastructure available to the KSEB scientists for pursuing 'better science' and by instigating the need for further scientific investigations with the help of better methods, Rastogi endeavoured to reinstate the lost public trust in science and invoked the possibility of achieving an authentic, objective, and accurate tech-nical solution to the problem. On the question of the epicentre, he carefully shifted the focus away from the Idukki dam. On the one hand, he argued that the epicentre was 30 kilometres away from Painavu in the south-western direction and proposed that Erattupetta was the epicentre.[52] This newly proposed epicentre was closer to Melukavu, the epicentre proposed by the KSEB scientists. On the other hand, he shifted attention away from the Idukki district itself while suggesting Palakkad as the most earthquake prone area in Kerala.[53] These strategies in effect helped him neutralize the RIS argument against the Idukki dam. The intervention of Rastogi thereby actually helped the KSEB push its real concerns while seeking consensus at different levels. There is no supposition here that Dr Rastogi was deliberately playing the game in his individual capacity—as an actor, his views were carefully mounted in the negotiations in the scientific public sphere through the interplay of several actors and the active role of the media. The timely and nuanced intervention of the 'impartial third party scientist' from outside the region with greater epistemic and moral authority successfully enforced closure to the controversy and eventually the closure helped in transferring the focus of the deliberations on the seismic activation of the region to the Mullaperiyar water dispute between Kerala and Tamil Nadu. The closure of deliberations through the discursively constructed figure of the superior scientist should be read against the earlier interventions of the IMD scientists who had failed to attain a similar closure due to the interventions of the media that challenged their credibility. Dr Rastogi claimed his special epistemological status through his carefully calibrated and technically nuanced interventions that appealed to the media and the public.

Shifting the Focus: The Mullaperiyar Dam and Regional Politics

The Mullaperiyar dam issue is still considered a serious political controversy in south India in its own right; the emphasis on the dam in the scientific public sphere was part of a wider political dispute between Kerala and Tamil Nadu on the question of sharing of water. The public controversy over the micro-earthquakes gradually became linked with the long-standing interstate political imbroglio of sharing the water of the Mullaperiyar dam and this shifting of grounds was achieved, as we have seen, through the public deliberations on the scientific interpretations of the incidents. The controversy began around the experts' contradictory calculations of the epicentre and magnitude. Gradually, the focus was shifted to the fear of the failure of the Idukki dam due to seismic activity at its vicinity. This was further shifted to a debate on the large number of dams in the Idukki district as causing increased seismicity in the region. Finally it was concluded that the epicentre was not the site of Idukki dam, further shifting the grounds of the debate to the oldest dam in the region as posing the risk of breakage due to earthquakes. This linked the earthquake controversy to the long-standing geopolitical dispute over the water reservoir. The immediate public consensus achieved over the vulnerability of the Mullaperiyar dam deserves our special attention as it reveals the coupling of regional politics with science.

The history of the Mullaperiyar dam goes back to the colonial period, as the gravity dam was built in 1896 using lime and mortar with the help of the British Army Engineering Corps under the leadership of Colonel John Penny Quick, the chief engineer of the project.[54] The dam was built as a result of an agreement (the Periyar Lease Deed) between the maharaja of the princely state of Travancore and the Madras government in 1886 to divert water from the Periyar River partly to the rain-shadow districts of the neighbouring Madras province.[55] The agreement was to share water on lease for 999 years. The pact continued in that form for a long time and was even renewed in 1970 unconditionally, but uneasiness about this agreement with Tamil Nadu developed into a major controversy since the building of the Idukki dam in the mid-1970s, 30 kilometres downstream. The new dam was conceived as a hydroelectric power project to provide electricity to Kerala, and the expectation was that the catchment area would be filled during the monsoon months. However, as time passed, the dam failed to fill up and it affected power generation. It became a necessity at this juncture to allow more water to reach the Idukki dam and the Mullaperiyar dam upstream turned to be a hindrance for the same. This has led to a long-lasting legal dispute between Tamil Nadu and Kerala regarding the treaty on the Mullaperiyar dam. Kerala wanted the storage level of the dam to be brought down to 136 feet and the Tamil Nadu government agreed to the same in 1979. Since then, there has been much tug of war between Kerala and Tamil Nadu about raising the height of water storage in the Mullaperiyar dam and the controversy has acquired several new dimensions of late. In 1998, the Government of Tamil Nadu filed a petition in the Supreme Court of India seeking to raise the water level to 152 feet. Kerala argued against this plea on security grounds, pointing out that the dam was more than a hundred years old. The issue, however, grew out of hand as a serious geopolitical controversy between the states only after the activation of the scientific public sphere in the wake of the micro-earthquakes.[56] That is to say that the scientists of the region and their expert interpretations has been central to the construction of the catastrophe discourse that focused on the possible failure of Mullaperiyar dam in more recent times.

Even though the public deliberations on the earthquakes veered towards the legal and political dispute over the Mullaperiyar dam only in a later phase, there had been a strong endeavour to link the dam with natural disasters to seek legitimacy to the arguments of Kerala, for the water dispute with Tamil Nadu was already a losing battle for the state.[57] Therefore, the debate over the epicentre and intensity of the tremors opened up a radically new possibility to regain ground for the battle and to forge a strong coalition of different actors towards defending the interests of the state. The regional press, scientists, and the state government worked together towards highlighting the potential risk created by the age-old dam (which was generally perceived as of no direct use to Kerala by then, but benefiting Tamil Nadu) for the people of Kerala in the context of seismic activation of the zone. Many of the news reports that appeared on the very first day of the earthquake controversy itself had attempted to highlight the threat of the earthquake to the dam to counter the Tamil Nadu government's continuous demand to raise its height. *Deshabhimani* argued that the Mullaperiyar dam was situated in the Idamalayar zone, a seismic zone that was very active, basing their claim on the authority of the CESS scientists.[58] The newspaper also contended that no tectonic study was undertaken until then in the case of this reservoir, and any attempt to raise the water level without a scientific study could prove disastrous.[59] Soon the issue became a central matter for the deliberations.[60]

Though it was the KSEB and the CESS scientists who proposed the link between the Mullaperiyar reservoir and the risk of seismicity in the beginning, in the following days other research teams also joined them in support of the possibility of a dam failure. The expert team of the Kerala government visited the dam site on 15 December and discovered cracks on the dam and suggested that it could be due to the earthquake.[61] The team was led by Lalitha Mickle, the then chief engineer of the Irrigation Dam Design Wing of the Irrigation Department of Kerala. The team suggested that even the permitted water storage

level of 136 feet was not safe and more scientific investigations to assess
the vulnerability of the dam and also the area were proposed.[62] The
investigative team of experts also announced a detailed scientific study
with the technical support of the IMD.[63] As we have seen earlier as well,
regional scientists were keen on establishing research coalitions with
national institutes, and this was crucial for safeguarding the interest of
Kerala and to outweigh similar (but more effective) attempts made by
Tamil Nadu and its scientists. The DMG also endorsed the vulnerability
of the dam in their scientific report to the state government.[64] They
discovered fresh cracks on the dam and suggested more collaborative
research with other scientific institutions.[65] They also pointed out that
the dam area is earthquake prone and demanded the establishment of
a seismic station nearby to conduct further investigations.[66] The scientists from the NGRI also discovered a crack at the base of the dam.[67]
The CESS also rose to the occasion by slightly modifying their epicentre argument: they suggested that the Mullaperiyar dam was within 50
kilometres of the epicentre of the earthquake.[68]

Taking note of the gravity of the issue, members of parliament (MPs)
from Kerala urged the union government to send a national expert team
to visit the dam site.[69] The newspapers raised an alarm about Tamil
Nadu's attempt to seal the cracks at this juncture.[70] The Tamil Nadu
government did not remain silent: the Dravida Munnetra Kazhakam
(DMK)[71] under the leadership of Dorai Murukan, the then minister of
the Public Works Department of Tamil Nadu, strongly denied Kerala's
claim that the dam was vulnerable.[72] To the dismay of Kerala, the
expert team of engineers appointed under the directive of the Supreme
Court to look into the matter adopted a stand in favour of the Tamil
Nadu government at this stage, declaring that the dam was safe.[73]
M.K. Parameswaran, the official representative of the Kerala government
in the commission opposed the majority opinion of the commission[74]
and demanded more detailed studies on the dam.[75] When the situation turned acrimonious, the Kerala government assigned an expert
committee for further investigation, with Dr M. Baba of the CESS as
the coordinator.[76] Interestingly, as mentioned earlier, the Mullaperiyar
dam had been a site of political and legal conflict between the regions,
wherein Tamil Nadu was always victorious.[77] But the translation of the
whole dispute into the language of science through a long process of
deliberation in the wake of the tremors helped Kerala to articulate its

demands in a more legitimate and persuasive way. It is also notable that the government and the scientists were in a symbiotic relationship—when the scientists helped the state to legitimize its arguments, their association with the government helped the scientists to handle the scientific controversy in a more conducive manner. The regionalist sentiments found support in the scientific and engineering community of Kerala. As the developments show, the negotiations in the scientific public sphere—since the tremors—gradually evolved towards a single point wherein a whole array of interests successfully converged on the Mullaperiyar dam. The dam turned into an 'obligatory passage point' for all the actors in the scientific public sphere (see Callon 1986; Latour 1987, 1988). That is, all those who were involved in the deliberations articulated themselves with reference to the threat created by the dam.

This alliance of the actors around the question of the dam hardened further when a second earthquake struck Kerala on 7 January 2001. This time both the KSEB and the CESS scientists had approximately the same judgement. The scientific and political lessons learned from their engagement in the earthquake controversy might have helped the scientists to localize the phenomenon more accurately, in a more consensual manner. Scientists from the CESS suggested Nadakkal, a place between Erattupetta and Theekkoi in the Kottayam district, as the epicentre.[78] The magnitude for them was 4.8 on the Richter scale. They also pointed out that the epicentre of the earthquake on 12 December also was at the same place, and, therefore, the latter was an aftershock of the former.[79] The KSEB scientists suggested that the earthquake was of an intensity of 5 and that the epicentre was near Peerumedu.[80] Although they differed on the location of the epicentre, for both groups of experts, it was at a safe distance away from the Idukki dam and closer to the Mullaperiyar dam. The intensity was approximately 5, strong enough to have an impact on the Mullaperiyar dam, but not adequate for the Idukki dam. Experts from other institutions also more or less corroborated these findings.[81]

The public and scientific consensus over the seismic vulnerability of the Mullaperiyar dam became possible in a situation where the battle lines had been drawn in terms of regional interests. The new earthquake that was localized closer to the Mullaperiyar dam augmented the demands of Kerala, and the tug of war between the regional governments continued. The Regional press went on to play its active role in the ongoing dispute by strongly supporting the Kerala government. Most of the newspapers reported deepening of cracks in the dam structure in the wake of the earthquake.[82] *Mathrubhumi* proposed the construction of a new dam in Mullaperiyar to replace the old, vulnerable one, which eventually has become the official standpoint of the government.[83] *Deshabhimani* quoted Dr S.K. Srivastava, the then additional director general of the IMD, as well as Dr K.R. Gopalakrishnan of the KSEB research wing, to suggest that the earthquakes had a lethal impact on the dam as the epicentre was near the dam site.[84] The newspaper also demanded reconsidering the decision taken by the expert committee in favour of Tamil Nadu to raise the water level to 142 feet. In an editorial on the issue published a little later, the spectre of the RIS theory rose again and the scientific authority of Dr Rastogi was cited to buttress the argument.[85] However, the experts were triumphant at this juncture in twisting the theory in favour of 'regional' interest. The DMG of the state government came out at this point with the argument that the Mullaperiyar dam was situated on one of the most unstable fault lines in the region.[86] In their report, the department urged the state government to take necessary steps to ensure the safety of the dam,[87] arguing that the epicentre of the earthquake on 7 January was at Kondoor, a site that was closer to the Mullaperiyar dam than was the epicentre of the first earthquake. Based on this assessment, they opined that the epicentre was gradually moving towards the dam, tremor after tremor. The expert committee appointed by the state government expressed the same opinion in its interim report.[88] Although the RIS theory was brought in by the newspapers to argue that reservoirs in the Idukki district (that included the Idukki dam) might be the cause of the tremors, now the argument was reversed to contend that the increased seismicity might cause the catastrophic collapse of the old and structurally vulnerable dam at Mullaperiyar. Such twists were common during other public controversies too; the same resources could be employed differently by different actors to win over the dispute. The success of the move was dependent on how it plugged into or triggered wider alliances and opinion formations.

On the other hand, the Central Soil and Material Research Station (CSMRS), New Delhi, a national research institute, appeared to rule in favour of Tamil Nadu in their investigation of the water pressure exerted on the dam structure. Their recommendations to the committee appointed by the Ministry of Water Resources of the central government even included raising the water level of the dam to 142 feet.[89] Although the Kerala representative in the committee objected to this recommendation,[90] the committee's final decision was in support of Tamil Nadu.[91] However, the articulation of the risk concerns of Kerala through technical language radically helped it to dissipate the previous consensus at the national level in favour of Tamil Nadu, and the committee's decision was in no way the final verdict on the matter. The perception of risk in Tamil Nadu was radically opposite: for them, the proposed decommissioning of the dam would be disastrous for the agriculture of four districts which have been completely dependent on the water from the Mullaperiyar dam (Pillai 2012). The strategy of Kerala was to present the catastrophe discourse that highlighted the looming disaster of dam failure with the help of scientific discourse, which seemed to be rewarding in the interstate dispute. Thereafter, Kerala began strongly demanding decommissioning of the Mullaperiyar dam and construction of a new one, which is now generally perceived as a possible way of resolving the dispute between the states.

The Looming Disaster

In spite of the agreement in the scientific public sphere on the threat posed by the Mullaperiyar dam, the scientific controversy that originated in the wake of the first earthquake of 12 December 2000 did not come to an end, although partial closures were achieved at different stages on specific moot points. The tremor that struck the region on 7 January 2001 reinvigorated the deliberations and the public anxiety about an impending catastrophe. The situation was aggravated when a major earthquake occurred in Gujarat on Friday, 26 January 2001, which the media and the scientific-citizen public perceived as confirmation of their apprehension about the local tremors as precursors to an impending disaster, and hence a failure of the earth scientists belonging to the state-owned departments and institutes who always denied such connections, as we shall see later in the chapter.

The second micro-earthquake that struck Kerala on 7 January 2001 (just a few weeks before the Gujarat earthquake) raised the question of the seismic vulnerability of the region as a new risk. *Malayala Manorama* and *Kerala Kaumudi* reported that the scientific belief that the Indian subcontinent was less prone to big earthquakes had been proven wrong, for the recent studies on the tectonic structure of the zone revealed the presence of fault lines.[92] Dr Srivastava of the IMD also endorsed the view that the tremors seemingly indicated the appearance of new fault lines in the Idukki zone.[93] *Mathrubhumi* contented that the second earthquake contradicted the previous scientific explanation that strong tremors of the same magnitude would not occur for a second time in Kerala, pointing to the discovery of fault lines.[94] The newspaper quoted Dr Rastogi as having suggested that new tremors might occur in the wake of the second earthquake.[95] *Deshabhimani* opined that the recurring tremors were an indication of the presence of a fault line that passed through the Idukki district and signalled the possibility of another earthquake.[96] *Madhyamam* evoked the RIS argument quoting the DMG as we discussed earlier,[97] and *Kerala Kaumudi* invited public attention to the construction of several dams in Idukki district as a major cause of the recurrent tremors in the area.[98]

In the subsequent days, lineaments (*bhoovillalukal*) and fault lines (*bhramsa rekhakal*) became a key topic of discussion. The fault line theory received much attention in the debate, and differences of opinion amongst the scientists concerning the possibility of a major earthquake in the region surfaced. The erosion of public trust in scientists was aggravated by the occurrence of repeated localized tremors in many parts of the region. Experts continued arguing that those were nothing but 'aftershocks' and negated the possibility of a major earthquake in the region.[99] This argument was defied in the scientific public sphere by employing three different strategies. The first involved a direct attack on the incapability and callousness of scientists in Kerala. The 'Letters to the Editor' column in newspapers turned out to be a crucial site for criticism, as we have also observed in the clinical trials controversy. A reader argued that the public did not trust the experts' denial of the possibility of any major earthquake in the region.[100] He suggested more research on the tectonic activities in the region in order to develop new techniques that help predict earthquakes. In a letter from a reader that appeared in *Mathrubhumi*,[101] the reader introduced himself as a scientist who had commenced his career at the CESS and attacked the researchers at the CESS for their neglect of and incompetence in studying the tectonic structure of the region.[102] According to the reader, the CESS possessed the basic modern scientific infrastructure to conduct effective research in earth sciences but its scientists lacked research experience. He added that Kerala was located in a geographical region with several active fault lines due to the presence of the Western Ghats on one side and the Arabian Sea on the other.[103] Consequently, Kerala was always prone to earthquakes but the scientists at the CESS never accorded serious attention to this fact.[104] Similar letters appeared in other newspapers too, challenging the experts' claims.[105]

Second, the newspapers began questioning the argument of the scientists that earthquakes could not be predicted. *Mathrubhumi* suggested that the foreshocks, development of cracks on the rocks, presence of certain minerals and radon gas in the ground water, colour change and rise of temperature, sudden rise and fall of water levels, and so on, in the wells could serve as indicators of earthquakes.[106] The news report suggested that geomagnetic changes and peculiar behaviours of animals and birds that occur before an earthquake are well-known earthquake warnings. *Deshabhimani* reported cases of successful earthquake prediction made by Chinese scientists who studied changes in animal and bird behaviour prior to earthquakes.[107] Both the news reports emphasized the failure of experts in many instances to take into consideration the warnings of the local public regarding the micro-changes in their area just before an earthquake. It was also observed that the Chinese and the Japanese scientists were making considerable advancements in the area of earthquake prediction, greater than Western experts.[108] The strategic invocation of the achievements of scientists belonging to non-Western cultural contexts ('Asian'/'oriental'/'ancient') hinted at the need for the regional scientists to turn towards scientists from these cultures (Chinese, for instance) for methodological insights. It also implied the limitations of Western science and the need for researchers in non-Western social contexts to be attentive to their own cultural resources (as the Chinese scientists did) to solve problems which have more local relevance. Thus, science was conceived as epistemologically rooted in multiple cultures and the regional earth scientists were blamed for being insensitive to the multicultural nature of modern science.

The third strategy involved a conscious effort on the part of the press to report incidents of micro-tremors in Kerala as well as earthquakes from all over the world to indicate that the regional tremors were not isolated events. The newspapers passionately reported even very small episodes of tectonic vibrations and related phenomena from every nook and corner of the state.[109] Comments from experts also were included in these reports. This tactic was intended to demonstrate that the tremors could not be dismissed as mere aftershocks, but were fresh phenomenon that deserved serious scientific attention. All the newspapers reported an earthquake in El Salvador in the Central America that claimed the lives of hundreds of people.[110] Such detailed reporting of tremors and earthquakes from all over the world gestured towards the experts' neglect of the tacit connection between earthquakes and the tectonic activity in Kerala.[111]

While this was the general mood of media reportage, scientists generally adopted a strong posture in insisting that there was no chance for a major earthquake in the region. But the totally unexpected and massive earthquake of magnitude 6.9 on the Richter scale that killed more than 17,000 people in Gujarat on Friday, 26 January 2001 drastically wore down the public's trust on the earth scientists of Kerala.

With this massive disaster, the public controversy concerning earthquake prediction attained new heights in Kerala. The Gujarat earthquake provided an opportunity to discuss larger issues about the tectonic structure of the Indian peninsula in relation to the tremors in Kerala as well as the scientifically unexplained connections between local tremors and earthquakes reported from other parts of the world. The Malayalam press widely and passionately covered the catastrophe in Gujarat. Interestingly, all the newspapers linked the event to the ongoing debates on earthquakes in Kerala. For instance, the point of difference between the IMD and the American Geological Survey over the magnitude of the Gujarat earthquake bore striking similarities with the controversy between the CESS and the KSEB, and the style of reporting of the dispute in the regional media was deeply influenced by the latter.[112] Similarly, in its editorial, *Mathrubhumi* discussed the possible connections of the disaster with the earthquakes that occurred in El Salvador and Japan just a few days previously.[113] The occurrence of low-intensity tremors one month prior to the Gujarat earthquake was pointed out and the manner in which scientists ignored local tremors was criticized in the regional press.[114] Many of the Malayalam newspapers discussed earlier predictions about an impending earthquake in Gujarat.[115] The recurrent instances of tremors in Kerala were alluded to be of a similar nature, hinting at the possibility of an imminent disaster in the region.[116] *Mathrubhumi* quoted Dr Harsh Gupta emphasizing the possibility of a major earthquake in Kerala in the future, contrary to the statement of the CESS scientists that there was no causal link between the Gujarat earthquake and the micro-tremors in Kerala.[117] The RIS argument also reappeared, and the pros and cons were deliberated.[118] In its previously mentioned editorial, *Mathrubhumi* suggested that large reservoirs like the Sardar Sarovar and Tehri Garhwal possibly had induced the quakes in Gujarat. A reader, in his letters to *Mathrubhumi* and *Kerala Kaumudi*, critiqued scientists for discarding the possibility of prediction and demanded more sincere and committed research from Indian scientists in this direction.[119]

Rumours of a likely earthquake circulated widely and created panic while the scientists kept on denying the likelihood of the same.[120] The newspapers never failed to bring in voices that went against the spirit of the proclamations of the scientific community.[121] Clashes among the regional scientific community also transpired into public at this stage. The earlier tensions between the scientists of the KSEB and the CESS reappeared at this juncture. K.R. Gopalakrishnan, the deputy director of the research wing of the KSEB, opined that an earthquake of magnitude 5.7 could occur in Kerala at any time, challenging Dr Kusala Rajendran of the CESS, who had categorically stated that possibility of a major earthquake in the region was negligible.[122] Gopalakrishnan offered scientific reasons in support of his argument and accorded it the imprimatur of scientificity by quoting Dr Rastogi. He also alleged that the CESS failed to coordinate the seismological research in Kerala, giving a peek into institutional politics within regional science. In short, the newspapers were attempting to voice the public anxiety about the possibility of a major disaster in Kerala similar to the one that struck Gujarat and the lack of trustworthiness of regional scientists who were not willing to pay heed to the public's perception of risk. And this also revealed the internal conflicts and institutional tensions among the regional scientific community.

Several fringe theories were mooted at this stage in public, thanks to a loss of trust in the official scientific explanations. These alternative propositions were quasi-scientific, as they lacked legitimacy among the scientific community. The scientific public sphere turned out to be the appropriate site to push forward such theories which negotiated the boundaries between science and non-science (see Chapter 6). A group of 'environmental scientists' raised the bogey of deforestation as a possible cause for the increase in the number of earthquakes in the state.[123] The 'sunspot theory' of Dr Mukherjee[124] that had been reported in the media just before the Gujarat disaster acquired more public attention at this stage. His theory became widely acknowledged and appreciated since his reportedly successful forecasting of the Gujarat earthquake through the analysis of changes in sunspot activity.[125] Based on recent photographs of sunspots from the National Aeronautics and Space Administration (NASA), the USA, more earthquakes in different parts of India in near future were supposedly predicted by him in the wake of the Gujarat earthquake.[126] Dr B.S. Bhattacharya of the IMD proposed that an effective earthquake warning system could be developed based on the detection of heightened tectonic activity just before the earthquake.[127]

The claims of some self-trained 'scientists' also were reported in the press, blurring the boundaries between scientific and popular cultures and introducing an 'epistemological cross-space' between them (Levina 2009: 109). For instance, *Madhyamam* reported the 'electromagnetic theory' suggested by P.N. Nair who had been studying the emission of 'electrons' in the atmosphere from the earth's core during periods of increased tectonic movements.[128] He named the theory after him as 'Nair Effect',[129] and claimed that several earthquakes including the recent ones in Kerala and Gujarat had been successfully predicted by him. Earthquake forecasts were made with the help of the 'micro shockwave detector', a device he had developed. Many of the claimants of similar theories, interpretations, and instruments presented themselves as 'experts' on earthquake prediction.

The 'Letters to the Editor' columns became a significant place for such articulations. For instance, a reader emphasized the lunar influence on tectonic activity.[130] He analysed the relationship between tremors in Kerala and the occurrence of full moon or new moon days for the last one decade and found that 80 per cent of the tremors appeared around full moon or new moon days. Another reader proposed a causal relationship between experimental nuclear explosions and earthquakes. He propounded the idea that the shockwaves from underground nuclear explosions had a tremendous impact on the stability of the tectonic plates that eventually triggered earthquakes since the shockwaves reverberate long after the quake and then get concentrated on a particular point under the surface of the earth.[131] It was the nuclear experiments of India and Pakistan that he claimed to be causative of the recurrent tremors and earthquakes in the Indian peninsula.[132]

Astrological predictions of earthquakes were also aired, albeit to a smaller extent. According to one astrologer, the Gujarat earthquake occurred on the most inauspicious day of the year: on that day the heavenly bodies were in a unique alignment, having an immense impact on the earth.[133] With reference to the rumour that one astrologer had been arrested in Gujarat for forecasting a fresh earthquake and released later when the prediction turned out to be true, a reader argued that this incident indicated the usefulness of astrology for earthquake prediction and insisted that more 'research' be done on the subject.[134]

The mushrooming of alternative explanations and technical strategies to forecast earthquakes in the public domain at this stage indicated a complete decline of public credibility of earth science and a loss of trust in regional experts. The fringe theories and explanations had been refuted by experts, but their voices were largely sidelined because of the erosion of public trust in experts in general.[135] The great importance arrogated to 'alternative science' in the scientific public sphere indicated that the loss of trust in science created conditions for giving equal credence to the 'scientific' claims of a multitude of actors. These alternative interpretations seemingly proliferated when expert advice was felt to be inadequate to deal with the risks or incompatible with the public perceptions of risk. The scientific public sphere generated a wider spectrum of ideas, interpretations, and expertise than what science had to offer. However, this did not mean that all the alternative explanations and expertise enjoyed the same status in the scientific public sphere. For instance, as we have seen, the astrological postulations enjoyed very little attention. This could be due to the peculiar formation of the civic epistemology of the region: astrology had already lost its sheen as a legitimate knowledge practice that could inform public debates on science.[136] As in the present case, astrological predictions were understood as irrational, whereas alternative explanations and fringe theories were accepted more readily by the public. Alternative explanations seemed to enter the scientific public sphere only if they used the language of science. Theories in other languages were largely rejected even during the collapse of public trust in experts and scientific institutions. This indicates that while the scientific public sphere redefined the boundaries of science, it also contoured new boundaries between science and non-science, even though these demarcations were more flexible and porous than that of the scientists.[137]

Unnatural Geophysical Events Strike Kerala

As mentioned earlier, localized tremors and associated phenomena began to be reported by media when the debate concerning the threat of an impending disastrous earthquake in Kerala was doing the rounds in the scientific public sphere. The public and the media generally perceived these events as implicitly connected with the seismic activation of the region and as signs of an ominous catastrophe. In spite of newspaper reports on oscillating water levels (*thirayilakkam*) in the traditional wells in various parts of the state following the Gujarat earthquake, scientists did not pay much attention to the phenomenon, and they had their own scientifically valid reasons for the same.[138] This created conflicts with local, mini-publics during experts' field visits, and the media highlighted these encounters. When an incident of oscillation of water in wells was noticed in Kalamassery, near Ernakulam, experts from the DMG visited the place, but the local public later complained that the scientists trivialized the issue and did not take their concerns seriously.[139] In another incident that happened a few months later in connection with the bizarre occurrence of fuming of hills in Idinjimala, near Kattappana in the Idukki district, the villagers demanded scientific investigations and an expert team from the CESS visited the place.[140] The experts could not douse the panic caused by the phenomenon and a dispute began after their visit.[141] Based on the expert advice the district collector stated that it was not fumes, but 'mist'.[142] The villagers challenged the scientists' explanation, pointing out that there was no 'fuming' of the hill when they visited the spot, and alleged that the expert advice was not based on any detailed investigation of the phenomenon.[143] The media also got a bashing from the local public for not independently assessing the expert opinion, and it is interesting to note that the newspapers reported this criticism of the media too.[144] Media reporting of local confrontations between citizens and scientists linked the local mini-publics with the scientific public sphere. Such incidents reported from micro-spaces of everyday interaction hint at the escalation of mutual distrust between the experts and the citizens and the pervasiveness of the collapse of the credibility of the regional scientific institutions.

The CESS scientists consoled the public at this juncture by explaining that localized tremors and water oscillation were due to 'seismic seizures', a harmless and common phenomenon usually occurring in the post-earthquake period.[145] They argued that it appeared in Kerala because of the Gujarat earthquake, but was innocuous and waning. On the contrary, the frequency of the occurrence of localized tremors gradually increased in Kerala and more unusual incidents came to light, such as the oscillation of water levels in wells and fissures on the walls of buildings. The regional press painstakingly continued reporting these incidents from all parts of the region.[146] Newspapers neatly knitted together incidents of earthquakes and volcanic eruptions from all over the world with the events of regional relevance and, hence, by amplifying the phenomena, tried to challenge the experts' view of the local incidents as inconsequential.[147]

The newspapers strongly critiqued the belittling of the incidents by the earth scientists. In its editorial on 15 February 2001, *Mathrubhumi* pointed out that the public anxiety created by the phenomenon was justifiable since the ongoing low-intensity tremors and water oscillation in wells indicated radical geological transformations.[148] The editorial problematized the experts' claim that these were 'low-intensity localized tremors' by showing that the phenomenon appeared across the region, and that the recurrent tremors, notwithstanding their low intensities, could weaken Kerala's geological make-up and hence the phenomenon needed to be scientifically studied. The newspaper also raised the question of RIS being behind the phenomena, trying to plug the issue into an earlier debate. In his response to the editorial, a reader highlighted the contradictions in the experts' arguments and criticized the geologists for trivializing the tremors as mere 'aftershocks' of the Gujarat earthquake, despite earth science having already proven that the 'foreshocks' were a valid indication of a massive earthquake in the near future.[149] He buttressed his argument by highlighting the incidence of water oscillation that had begun in Kerala much before the Gujarat quake. The newspaper went a step further when it launched a special essay-series written by Sasidharan Mangathil, a geologist-turned-journalist.[150] The essays meticulously challenged many of the scientific explanations, and pointed out that the public had lost its trust in scientists.[151] The author strongly propounded the activation of fault lines in the region as the reason for the continuing tremors.[152] The active involvement of scientifically well-informed journalists with domain knowledge in relevant fields helped the newspapers enjoy greater public credibility than the scientists themselves. These journalists skilfully addressed the public anxieties and their concerns about the risks raised by the unnatural geophysical phenomena: this was in sharp contrast to the failure of the scientists in engaging with the scientific-citizen public. While the experts' interaction with the public was framed by the deficit model,[153] these learned journalists triumphantly demonstrated a better communication model that took the public in confidence as partners in dialogue. Although Malayalam newspapers lacked special science desks, many of them encouraged journalists with domain knowledge in relevant scientific disciplines to engage with scientists and published their special reports.[154]

Vanishing Wells

The controversy received a fresh lease of life when wells started collapsing in Kerala in the month of March 2001. The regional press often named the event as 'well vanishing' (*kinar aprathyakshamakal*) since they got filled with mud due to the collapse of their walls.[155] These were the traditional wells that each household had, and which provided for the water requirement of the family. The traditional wells are a major source of water in Kerala and hence directly linked to the everyday life of the people. A major anxiety for the public in this context was the phenomenon's relation to the regional tremors as well as the Gujarat earthquake.[156] According to one news report, 123 wells had collapsed in the region till 12 July 2001.[157] From July onwards the number of incidents reported in the media increased exponentially, and collapsing wells became very common in Kerala. Several other unusual geological phenomena too were reported along with the increasing number of well collapses. These included the appearance of cracks on the walls of buildings, ground fissures, leaf fall, tunnel formation, change in the colour of water as well as bubbling and boiling of water in wells, ponds, and paddy fields, and coloured rains.[158] The unleashing of such a large spectrum of unusual geological incidents in the region created new problems for the scientists, because the local public demanded their visit to localities where such incidents had occurred in order to provide technical advice and thereby dispelling their anxieties and satiating their scientific curiosity. As we have seen in other contexts too, in several instances, the local public entered into heated arguments with the experts, challenging the scientific explanations being offered.[159]

Public criticism of the inactivity of the government and the lack of legitimate and convincing scientific interpretation of the phenomena forced the state administration to intervene in the controversy. A research team appointed by the Kerala government studied 50 cases of well collapses reported from different parts of the region. The research team constituted under the directive of the then CM of Kerala (A.K. Antony) consisted of scientists from the CESS, the Kerala State Groundwater Department (KSGD), the Central Groundwater Board, the Mining and Geology Department, and the GSI. The expert team pointed out in their interim report that the phenomenon was not earthquake related.[160] For them, the phenomenon was precipitated by extreme pressure being exerted on the walls of the wells due to the accumulation of ground water during summer rains (April–May) and when the torrential monsoon rains hit the region in June and it was mainly the 'poor construction of the wells' that had to be blamed.[161]

In sharp contrast to this report's conclusions, the preliminary inferences from an ongoing study carried out by a team of scientists from the KSGD propounded that the tectonic vibrations on the earth's crust after the Gujarat earthquake, along with groundwater pressure, was the reason behind the well collapses.[162] The study concentrated on the collapsed wells at Pavangadu in Kozhikode district from where the maximum number of well collapses had been reported. The research team[163] did a detailed analysis of six collapsed wells in the locality, situated within a radius of 2 square kilometres.[164] This new study which established the linkage between well collapses and earthquakes created much debate in the following days, as it corroborated the apprehensions of the public. The government-appointed team of experts strongly opposed the findings of the new study by the KSGD researchers. The director of the DMG, Dr Padmanabhan Nair, staunchly questioned the credibility of the study; he also highlighted the participation of veteran scientists from the KSGD in the official team of experts, hinting that the new study was done by a team of young and inexperienced researchers from the KSGD, and that their study lacked institutional endorsement.[165] Similarly, Dr Baba, the director of the CESS, vociferously denounced the conclusions of the new study and eliminated the possibility for any connection between the earthquakes and the well collapses, stating that not a single tremor had been reported from the region when the well collapses occurred.[166] He reiterated the official team's estimation of the changes in rainfall pattern, 'unscientific' well construction, and soil characteristics as the causes of the phenomenon. Dr Baba further explained the water oscillation in wells as caused by 'micro-tremors', but asserted that the tremors had no relation with well collapses.[167] The director of the Seismology Division of the Bhabha Atomic Research Centre (BARC), Mumbai, Dr G. Jayachandran Nair, also dismissed any tacit connection between earthquakes and well collapses, suggesting that well collapses were due to 'instabilities' encountered just 30 metres below the earth's surface.[168]

The strong defence raised by the three directors of these scientific research institutions against the preliminary analysis from an independent study conducted by a team of young researchers, who lacked institutional support and epistemological power, was incredible but failed to influence public opinion. Albeit being hierarchically superior and politically powerful, the official team of experts faced strong opposition from the public at this stage, who explicitly supported the other team of researchers. Several questions were raised against the former's analysis. Apparently the official team of experts' analysis of the traditional wells as the culprit did not go well with common sense: these wells have been a part of everyday life in the region, and in most cases, the main source of water for the household and the neighbourhood. The traditional wells were built with the help of indigenous techniques based on a thorough knowledge of the locality, the nature of the soil, and the geological structure of the area. In this sense, they were built with substantial indigenous engineering skills supported by sound local knowledge. The collapse of these wells, hence, was an unprecedented phenomenon for the public and they naturally linked it with the tremors and other geophysical phenomena. The findings of the official team of experts, therefore, conflicted with public's risk perceptions and lay-knowledge. And the public found the study of the KSGD scientists more readily addressing their anxieties and risk perception.

While discussing the final report of the governmental expert team, *Malayala Manorama* highlighted the vagueness of their explanation regarding the nature and pervasiveness of groundwater pressure that had been portrayed as the cause of well collapses.[169] In its editorial on the same day, the newspaper drew attention to the linkage between earthquakes and well collapses, though they did not criticize the expert opinion.[170] *Mathrubhumi* reiterated its earlier standpoint that the tremors in the region were not a localized phenomenon and connected them to the well collapses.[171] For the newspaper, the recent phenomena indicated vital alterations in the geological structure of the region that made it vulnerable. Therefore, more in-depth scientific investigations were demanded. *Deshabhimani* in a report criticized the scientists for the ambiguities and contradictions in their statements as well as the callousness of the state government.[172] The news report also highlighted the differences of opinion between different scientific agencies who participated in the official team of experts. *Madhyamam* in its editorial brought attention to the fact that despite being hit by numerous rainy seasons witnessing heavy rainfall, well collapses had never been reported before.[173] The editorial underscored that the earthquakes in the region were the primary cause for the phenomenon and recommended more research on the subject. Such a detailed study of the geological structure of Kerala was important, the editorial argued, for drawing up a better environmental as well as developmental policy for the state. Many readers also expressed similar concerns, turning down the technical assessment of the official team.[174]

The debate took a novel turn at this point, thanks to the intervention of and the competition between the two leading newspapers—*Malayala Manorama* and *Mathrubhumi*.[175] *Mathrubhumi* organized its reports and editorials around the 'fault line theory', which took its own course in the scientific public sphere a few months ago, with the essay series it had published on the earthquakes in the region.[176] On 27 July, the newspaper published a report that drew attention to the argument of some scientists that the fault lines in the region were getting activated.[177] The report also alerted the public to the fear among some scientists regarding the occurrence of strange geological phenomena as presaging a massive earthquake in the region in the near future. John Mathai, a scientist from the CESS, was quoted in the newspaper as suggesting that well collapses occurred in a particular direction running parallel to the coastal line.[178] He pointed out the existence of several fault lines in the same direction, and argued that the activation of these lines was causing the well collapses. Mathai interpreted the associated geological phenomena such as the water level oscillation in wells and the localized micro-earthquakes from this viewpoint.[179] *Mathrubhumi* published an editorial the next day, substantiating the fault line theory.[180] The final report of the study conducted by the KSGD scientists at Pavangadu was reported in the newspaper as further endorsing the fault line theory.[181] *Mathrubhumi* published a very long article by K.V. Mohanan, the coordinator of the KSGD research team, too.[182] *Deshabhimani* and *Madhyamam* also highlighted the study.[183] The fault line theory of well collapses was appreciated in the scientific public sphere as a valid scientific explanation following these news reports.[184]

The scientific endorsement of the fault line theory created great anxiety among the public regarding the changing tectonic structure of the region.[185] *Mathrubhumi's* interventions were crucial in establishing the fault line theory as a convincing scientific explanation to the events. The emergence of the newspaper as a prominent mediator between science and the scientific-citizen public alerted *Malayala Manorama*, their foremost competitor, to be more proactive. The latter thus went a step ahead, commissioning its own team of experts to undertake fresh research on well collapses.[186] The team consisted of two reputed scientists (Dr Janardhan G. Negi and Dr Arun Bapat) from outside the state of Kerala.[187] They spent three days in Kerala studying the phenomenon and consulting other researchers[188] and explained that well collapses in the region were the net result of geological and geo-hydrological processes.[189] The vanishing wells were conceived by them as a process in continuation with the earthquakes that struck the region in the months of December and January. Increase in 'pore pressure' as well as changes in the 'stress field' due to the pressure that had accumulated in the bedrocks after the earthquakes, they argued, led to changes in the course of ground water streams. When the rain stopped, there was a sudden decrease in the amount of groundwater, and the vacuum created under the well due to these complex processes, in their opinion, caused the wells to collapse. They also noted that the collapsed wells were situated at the nodal points where the fault lines intersected, and referred to the KSGD study as substantiating their argument.[190] The new argument blended both the groundwater pressure theory and the fault line theory within its fold, and the status of these as two mutually contesting and incommensurable hypotheses had been defused by the intervention of the expert team from outside Kerala with superior epistemological status.[191] More detailed investigations of the fault lines as well as the tectonic structure of the region were proposed by them.[192] In its editorial response to the findings of the research team, *Malayala Manorama* emphasized the need for more scientific research on the subject.[193] The competition between the rival newspapers, thus, turned to be creating new journalistic practices, indicating the functional differentiation of media in its interaction with science in the context of risk politics in Kerala. By commissioning its own independent team of experts to give technical advice (*Malayala Manorama*) or by strongly endorsing a particular scientific interpretation against the official interpretations (*Mathrubhumi*), media was redefining its role as an arbitrator and agenda setter in the public deliberations, and their proactive role in the controversy made the scientific-citizen public trust them more than the official expert teams appointed by the government.

However, other newspapers completely ignored the initiative of *Malayala Manorama*. *Mathrubhumi* continued pushing the fault line theory, laying more emphasis on the necessity of a comprehensive study on the lineaments and fault lines in the region.[194] The latter reiterated the possibility of an impending earthquake in the region and pointed out that the scientists who proposed the fault line theory could not do further research due to the lack of institutional infrastructure. Taking into cognizance the limitations of the regional research system, *Mathrubhumi* urged the state government to request national scientific institutions, which have better facilities and expertise, to conduct further research on the region.[195] As we have observed, the public credibility of the regional scientific establishment touched the bottom at the final lap of the controversy. However, this also helped some young researchers at its margins to emerge as successful interpreters, whom the public trusted due to their greater openness in engaging with the risk perceptions of the scientific-citizen public. There was, nevertheless, no permanent closure to the controversy, nor did the unusual geophysical phenomena cease to occur. Similar events continued intermittently in the region for a long time and so did the debates.[196]

***

The public controversy that began in the wake of the tectonic activation of Kerala with a series of micro-earthquakes, as we have seen, passed through many phases, with the foci of the deliberations shifting constantly. The scientific community of the region, especially the earth scientists, came under public criticism for their interpretations of the phenomena, and this brought the internal contradictions of science and its political dynamics into full public view. These contradictions became publicly known when scientific explanations to the unprecedented natural phenomenon that struck the region were sought by the public, and the regional press relied upon diverse sources of scientific information to satiate the demand. This led to the appearance of contradictory explanations in the media that catalysed public scepticism about the regional scientific institutions and the experts' scientific credentials. Gradually, the controversy intensified with a corresponding decline of public trust in science when the experts failed to address the risk-related anxieties of the public. The risk perception of the scientific-citizen public in the region was heavily dependent on historically developed cultural registers such as environmental crisis, public apprehensions about massive dam projects, and the active public memory of tremors in the region. Partial closure was obtained at each stage of the controversy either through the tactical interventions of experts from outside the region, who enjoyed higher public credibility, or when the interests of diverse actors converged, as in the case of the Mullaperiyar dam. Regional newspapers played a dynamic role as mediators between the experts and the scientific-citizen public and they showed great enthusiasm in representing the polyphony of opinions in the scientific public sphere.

The controversy had been generated by the regional press following the contradictions and ambivalences in expert interpretations of the earthquakes. Once the scientific public sphere was activated, scientists were forced to engage with the media and the public. They had to present their arguments and interpretations before the public and respond to questions and criticisms. Whenever the experts became adamant or failed to engage with the concerns and perceptions of the public, they faced massive loss of scientific credibility. The media's proactive role in bringing in scientific resources to the public helped the non-scientist actors to develop new interpretations against the moderate explanations extended by the experts who were constrained by institutional structures and the epistemological culture of science. The public controversy over earthquakes in Kerala, thus, opened up the internal mechanism of science to the public and exposed the connections it had with other social institutions that shaped the way knowledge was produced.

The case also showcases some unique characteristics of the mutual resonance of science and media. The newspapers seriously participated in the debate by patronizing certain scientists and institutions and in a unique occasion, hiring a team of experts. Scientists were forced to be accountable to the scientific-citizen public and their expert interpretations and advice often conflicted with the lay-knowledge and risk perceptions of the public, making the scientists and their institutions suffer from loss of trust and credibility. Under the pressure of retaining their readership, the newspapers also contributed to the process of public scrutiny of science by staging and shaping the risk controversy. The strangeness of the phenomena as well as their direct impact on the daily lives of the people facilitated the process of deliberation in the scientific public sphere, establishing a unique form of public engagement with science.

Notes

1. 'Bhoochalanathil Keralam Nadungi', *Malayala Manorama*, 13 December 2000, p. 1; 'Bhoochalana Kendram Melukavu: Karanam Dam Ennu Soochana', *Kerala Kaumudi*, 13 December 2000, p. 1.

2. Kerala has a long history of micro-tremors striking the region and the regional media always reported these incidents with great alarm, despite the experts' assertion that these tremors are harmless. This indicates that micro-earthquakes had already been discursively constructed as a source of great risk by the regional media.

3. The CESS is an autonomous research centre instituted by the Government of Kerala in 1978, and is situated in Akkulam, Thiruvananthapuram. The centre carries out research in different areas of earth system studies from a multidisciplinary perspective, including studies on paleomagnetism, natural hazards, coastal zone management, and use of resource studies for local planning. The centre is currently known as the National Centre for Earth Science Studies (NCESS), and is affiliated to the Ministry of Earth Sciences of the Government of India (GoI).

4. There are 14 dams in the district. These are Mullaperiyar, Kundala, Mattuppetti, Kallar, Kallarkutty, Neriyamangalam, Anayirankal, Ponmudi, Panniyar, Idamalayar, Idukki, Cheruthoni, Kulamavu, and Lower Periyar. See P.K. Prakash, 'Innale Nadanna Bhoochalanam, Oru Munnariyippu', *Madhyamam*,

13 December 2000, p. 4, editorial page article; 'Bhoochalanangalkku Karanam Anakkettukal', *Madhyamam*, 13 December 2000, p. 1.

5. 'Chilayidathu Veendum Neriya Bhoochalanam', *Mathrubhumi*, 14 December 2000, p. 1.

6. 'Keralam Bhookampa Sadhyatha Mekhala', *Madhyamam*, 14 December 2000, p. 5. The investigations of the DMG as reported in the newspaper clearly indicated the gradual activation of the region as earthquake prone.

7. 'Bhoochalana Kendram Kandethan Sramam Arambhichu', *Malayala Manorama*, 15 December 2000, p. 1.

8. The expert team had B.S. Rane, Assistant Meteorologist, and M. Jayagopal, Senior Scientific Assistant, as members.

9. See the report of the investigations by the IMD. 'Bhoochalanam: Kendrabindu Melukavinum Painavinum Madhye Akanida', *Madhyamam*, 16 December 2000, p. 5.

10. 'Kendra Vidagdha Sangham Melukavumattavum Painavum Sandar-shichu', *Deshabhimani*, 15 December 2000, p. 1.

11. The IMD's close association with the KSEB was noted by *Kerala Kaumudi* and *Deshabhimani*, similar to the media's observations about the Purvesh Parikh Commission's tacit alliance with the RCC administration in the previous chapter. *Deshabhimani* reported that the expert team had been received at the airport by the KSEB research wing director K.R. Gopalakrishnan. *Kerala Kaumudi* reported that the expert team was invited by the KSEB. See 'Bhoochalanam Mullaperiyar Damil Villalukal Veezhthi', *Kerala Kaumudi*, 16 December 2000, p. 1; 'Bhoochalanam: Kendra Sangham Ethi, Pradhamika Padana Report Randazhchaykkullil', *Deshabhimani*, 15 December 2000, p. 5.

12. 'Vydyutha Boardinte Bhoochalana Kanakkeduppu Pali', *Madhyamam*, 14 December 2000, p. 1.

13. Hartal is a form of public strike practiced in South Asian countries. During hartals, people voluntarily close shops and offices, and stop vehicle movement to mourn or register protest. The practice has its origin in the Indian Independence Movement. In Kerala, there is a general public opinion formed in recent decades against frequent and coercive imposition of hartals by the political parties.

14. *Deshabhimani*, as mentioned elsewhere, is a newspaper published by the Kerala CPI(M), and the CITU (Centre of Indian Trade Unions) is the trade union of the same political party.

15. 'Bhoochalanam: Samaram Moolam Kanakkeduppu Paliyenna Pracha-ranam Avasthavam', *Deshabhimani*, 16 December 2000, p. 7.

16. 'Bhoochalanam: Samaram Moolam', *Deshabhimani*.

17. 'Bhoochalanam Valutho Cherutho Ennu Tharkkam', *Kerala Kaumudi*, 17 December 2000, p. 1. See also 'Keralathil Vyapakamaya Bhoochalanam', *Deshabhimani*, 13 December 2000, p. 1, lead article.

18. The team consisted of Dr Harendranath and Dr Balachandran. See 'Mullaperiyar Dam Bhookampa Meghalayil', *Malayala Manorama*, 17 December 2006, p. 1.

19. 'Bhoochalanam Valutho Cherutho', *Kerala Kaumudi*.

20. 'Bhoochalanam Valutho Cherutho', *Kerala Kaumudi*.

21. All the newspapers except *Deshabhimani* highlighted the issue on the first day itself. *Deshabhimani* seemingly kept quiet about the issue, as the CPI (M) led Left Democratic Front was ruling the state. B. Ajith Babu. 'Chalana Kendram Periyar-Idamalayar Bhramsa Meghala', *Malayala Manorama*, 13 December 2000, p. 8; 'Keralathil Bhoochalanam, Paribhranthi', *Mathrubhumi*, 13 December 2000, p. 1, lead article; 'Bhoochalanangalkku Karanam Anakkettukal', *Madhyamam*; 'Bhoochalana Kendram Melukavu', *Kerala Kaumudi*.

22. Idukki dam is still the highest arch dam in Asia (555 feet). Its power-house at Moolamattom is the longest underground power station in India. The project has two more dams at Cheruthoni and Kulamavu. Construction of the project was technically and financially supported by the Canadian government. For more details, see http://expert-eyes.org/deepak/idukki.html and http://idukki.nic.in/dam-hist.htm (accessed on 27 March 2016).

23. 'Chalana Kendram Periyar', *Malayala Manorama*.

24. 'Bhoochalana Kendram Melukavu', *Kerala Kaumudi*.

25. 'Iniyumundakan Sadhyatha', *Madhyamam*, 13 December 2000, p. 1.

26. See 'Iniyumundakan Sadhyatha', *Madhyamam*; Prakash, 'Innale Nadanna Bhoochalanam', *Madhyamam*; 'Bhoochalanangalkku Karanam Anakkettukal', *Madhyamam*.

27. See 'Iniyumundakan Sadhyatha', *Madhyamam*.

28. Apparently journalists asked the scientists difficult questions about the RIS with reference to the Idukki reservoir in press conferences and interviews.

29. 'Mullaperiyar Surakshithamalla, Idukkiykku Bheeshaniyilla', *Malayala Manorama*, 13 December 2000, p. 9; 'Mullaperiyar Dam, Uyaram Koottiyal Bhoochalanathinu Sadhyatha', *Deshabhimani*, 13 December 2000, p. 1; 'Bhoochalanangal Mullaperiyarinu Bheeshani', *Kerala Kaumudi*, 13 December 2000, p. 2.

30. The debate on the possible threat raised by the Mullaperiyar dam deserves more attention as it is an ongoing debate interlinked with the inter-state politics between Tamil Nadu and Kerala. The issue will be discussed in detail in the next section.

31. 'Bhoochalanam, Mullaperiyar Veendum Bheeshaniyakunnu', *Mathrubhumi*, 13 December 2000, p. 5; 'Mullaperiyar Dam, Uyaram', *Deshabhimani*; 'Bhoochalanangal Mullaperiyarinu Bheeshani', *Kerala Kaumudi*.

32. The assessment of the magnitude and epicentre of the earthquakes has been a regular official responsibility for both the governmental institutions.

33. 'Bhoochalanam Oru Munnariyippu', *Madhyamam*, 14 December 2000, p. 4, editorial.

34. The Koyna dam was built in Maharashtra in 1962, and a major earthquake of a magnitude of 7 occurred at the dam site (Koyna Nagar) on 11 December 1967. The earthquake partially destroyed the dam as cracks appeared on it. Many geologists believe that the earthquake was induced by the reservoir. For a detailed study, see Catone-Huber and Smith (n.d.).

35. Big dam projects like the Tehri and Vishnuprayag were located in the Himalayas, the editorial pointed out. See 'Bhoochalanam Oru Munnariyippu', *Madhyamam*, 14 December 2000, p. 4, editorial.

36. R. Sasi Sekhar, 'Anakkettukalum Bhoochalana Prerakam', *Mathrubhumi*, 17 December 2000, p. 4, editorial page article.

37. The construction of the Idukki dam began in 1969 and it was commissioned in 1976. The significance of 1974 was not explained in the news report, but it seems to be based on a technical paper authored by Dr B.K. Rastogi where he states that the filling of the reservoir started in 1974. See Rastogi (2001).

38. *Mathrubhumi* published a series of write-ups on RIS. It presented the statements of renowned environmental activists and public intellectuals on three consecutive days. On the first day, however, Arundhati Roy and Baba Amte's statements on RIS were published with an exclusive focus on the Mullaperiyar dam. See 'Mullaperiyar, Duritham Anubhaviykkuka Nammal Thanne: Arundhati, Amte', *Mathrubhumi*, 18 December 2000, p. 1. The newspaper broadened its focus on successive days, when Medha Patkar, Sundarlal Bahuguna, and Vandana Shiva were invited to make public statements on the controversy. See 'Keraleeyar Unaranam: Medha, Njan Ningalude Koode: Bahuguna', *Mathrubhumi*, 19 December 2000, p. 1; 'Keralathilethu Manushyar Undakkiya Bhookampam: Dr. Vandana Shiva', *Mathrubhumi*, 20 December 2000, p. 1.

39. See Chapter 3. Newspapers quite often used different 'discourse frames' to situate and emphasize issues and arguments. See Chapter 7 for a detailed discussion on this narrative strategy.

40. Rastogi turned to be a big player in attaining closure by 18 December. All newspapers have widely covered his statements, but even this was not going unopposed. *Mathrubhumi*, the newspaper that strongly supported RIS, simply ignored Rastogi. The newspaper published a small, one-column news item about his arrival on 17 December. See 'Bhoochalanam Padikkan Rastogi Ethi', *Mathrubhumi*, 17 December 2000, p. 13.

41. 'Valiya Bhookampam Undakan Sadhyatha Illa: Rastogi', *Malayala Manorama*, 17 December 2000, p. 1; 'Van Bhoochalanangalkku Keralathil Sadhyatha Illa: Rastogi', *Deshabhimani*, 18 December 2000, p. 1; 'Neriya Bhoochalanangal Thudarum', *Madhyamam*, 18 December 2000, p. 1.

42. 'Mullaperiyar Damil Vereyum Villalukal, Idukki Surakshitham', *Malayala Manorama*, 18 December 2000, p. 1; 'Urulpottal Meghalakalil Mazhakkalathu

Jagratha Venam: Dr. Rastogi', *Malayala Manorama*, 18 December 2000, p. 9; 'Bhoochalana Kendram Adukkam, Erattupetta, Poonjar Pradeshangalkku Idaykku: Dr. Rastogi', *Deshabhimani*, 19 December 2000, p. 12.

43. 'Mullaperiyar Damil Vereyum', *Malayala Manorama*.

44. 'Mullaperiyar Damil Vereyum', *Malayala Manorama*.

45. 'Mullaperiyar Damil Vereyum', *Malayala Manorama*; 'Bhoochalana Karanam Idukki Damalla: Rastogi', *Deshabhimani*, 20 December 2000, p. 6; 'Bhoochalanam, Bhayashanka Vendennu Vidagdhabhiprayam', *Kerala Kaumudi*, 18 December 2000, p. 1.

46. 'Bhoochalanathinu Karanam Jalasambharanikalalla, Vidagdha Report', *Malayala Manorama*, 24 December 2000, p. 9.

47. 'Van Bhoochalanangalkku Keralathil', *Deshabhimani*.

48. Bhoochalana Kendram Adukkam', *Deshabhimani*, p. 12; 'Keralathile Bhoochalanathinu 5 Muthal 8 Kilometre Vare Vegatha', *Madhyamam*, 19 December 2000, p. 10. 'Bhoochalana Karanam Idukki', *Deshabhimani*.

49. 'Keralathile Bhoochalanathinu 5 Muthal', *Madhyamam*.

50. 'Keralathile Bhoochalanathinu 5 Muthal', *Madhyamam*.

51. 'Keralathile Bhoochalanathinu 5 Muthal', *Madhyamam*.

52. 'Bhoochalana Kendram Adukkam', *Deshabhimani*; 'Bhoochalana Karanam Idukki', *Deshabhimani*. He has also published a short technical paper on the earthquake and its aftershock on 7 January 2001. In the paper he strongly refutes the role of the Idukki reservoir in inducing the earthquake, but concludes that '[o]ccurrence of several strong earthquakes [in Kerala] along known lineaments in a short span of time suggests that the ambient stress regime in the region has attained criticality in view of which the possible effect of dams may have to be studied more closely' (Rastogi 2001: 274).

53. 'Bhoochalana Karanam Idukki', *Deshabhimani*.

54. Its construction had begun in September 1887 and was completed in February 1896, although the inauguration was done one year before, in October 1895. The historical details given here are from Mangathil (2008).

55. The agreement was signed on 29 October 1886.

56. For more on the legal, political, and historical aspects of the Mullaperiyar water dispute between Kerala and Tamil Nadu, see Achyuthan (2012), Damodaran (2014), George (2011), Mangathil (2008), and Thomas (2012).

57. Even on the day before the first earthquake, there was an editorial page article in *Madhyamam* about the environmental destruction that the raising of the water storage level in the Mullaperiyar dam could cause. See Vinodkumar Damodar, 'Mullaperiyar, Kanathe Pokunna Paristhithi Prasnangal', *Madhyamam*, 12 December 2000, p. 4.

58. 'Mullaperiyar Dam: Uyaram Koottiyal Bhoochalanathinu Sadhyatha', *Deshabhimani*, 13 December 2000, p. 1. Although other newspapers shared the concern, they published similar reports only on the local pages.

59. 'Mullaperiyar Dam: Uyaram', *Deshabhimani*. See also *Mathrubhumi* and *Malayala Manorama* of 13 December 2000 for similar news reports.

60. Many of the newspapers published editorials on the issue. See 'Mulla-periyar Pattakkarar Enthinu Puthukkanam?', *Kerala Kaumudi*, 19 December 2000, p. 4; 'Mullaperiyar Damile Jalanirappu Thazhthanam', *Mathrubhumi*, 17 December 2000, p. 4; 'Mullaperiyar Damile Jalanirappu Thazhthan Vaikaruthu', *Mathrubhumi*, 19 December 2000, p. 4; 'Mullaperiyar, Durantham Vilichu Varuththaruthu', *Deshabhimani*, 18 December 2000, p. 4.

61. 'Mullaperiyar Daminu Villal, Bhookampam Moolamennu Samshayam', *Mathrubhumi*, 16 December 2000, p. 1, lead article; 'Bhoochalanam, Mulla-periyar Damil Villal', *Deshabhimani*, 16 December 2000, p. 1, lead article; 'Bhoochalanam Mullaperiyar Damil Villalukal Veezhthi, Kooduthal Padanam Venamennu Vidagdha Sangham', *Kerala Kaumudi*, 16 December 2000, p. 1; 'Mullaperiyar 136 Adi Sambharanavum Surakshithamalla, Vidagdhar', *Madhyamam*, 17 December 2000, p. 11; 'Mullaperiyar Daminu Villal Kandethi', *Malayala Manorama*, 16 December 2000, p. 16.

62. 'Mullaperiyar Daminu Villal', *Mathrubhumi*; 'Bhoochalanam, Mulla-periyar Damil', *Deshabhimani*.

63. 'Mullaperiyar Daminu Villal', *Mathrubhumi*.

64. 'Mullaperiyar Dam Bhookampa', *Malayala Manorama*; 'Mullaperiyar Villal Bhoochalanam Moolam Thanne', *Mathrubhumi*, 17 December 2000, p. 1, lead article; 'Bhoochalanam Neridan Ini Daminu Seshiyilla', *Deshabhimani*, 17 December 2000, p. 1; 'Mullaperiyar Dam Bhookampa Meghalayil', *Kerala Kaumudi*, 17 December 2000, p. 1; 'Mullaperiyar Bheethi Vithaykkunnu', *Kerala Kaumudi*, 17 December 2000, p. 1.

65. 'Bhoochalanam Neridan Ini', *Deshabhimani*.

66. 'Bhoochalanam Neridan Ini', *Deshabhimani*.

67. 'Mullaperiyar, Aditharayilum Villal Kandethi', *Madhyamam*, 20 December 2000, p. 1.

68. 'Bhoochalanam Neridan Ini', *Deshabhimani*.

69. 'Mullaperiyarileykku Unnathathala Kendra Sanghathe Ayaykkanam, MPmar', *Deshabhimani*, 19 December 2000, p. 6; 'Mullaperiyar Dam, Vidag-dha Sanghathe Ayaykkanamennu Keralam', *Mathrubhumi*, 19 December 2000, p. 5.

70. 'Mullaperiyar Surakshithamennna Pracharanavumayi DMK', *Desha-bhimani*, 19 December 2000, p. 12; 'Mullaperiyar Damil Shakthamaya Chorcha', *Deshabhimani*, 20 December 2000, p. 1; 'Mullaperiyar Chorcha Shakthamayi, Villaladaykkan Tamil Nadu Sramam', *Mathrubhumi*, 20 December 2000, p. 1; 'Mullaperiyar, Villalukalil Kummayam Poosaruthennu Keralathinte Avasyam', *Malayala Manorama*, 19 December 2000, p. 1; 'Mullaperiyar, Villalukalil Kummayam Poosal Thudangi', *Malayala Manorama*, 20 December 2000, p. 16.

71. The DMK is a major political party in Tamil Nadu.

72. 'Mullaperiyar Surakshithamennna Pracharanavumayi', *Deshabhimani;* 'Mullaperiyarine Badhichittilla', *Mathrubhumi,* 18 December 2000, p. 1; 'Mullaperiyar, Jalanirappu Uyarthan Tamizhnattil Veendum Prakshobham', *Madhyamam,* 19 December 2000, p. 10.

73. Dr B.K. Mittal, the chairperson of the committee, was a dam engineering expert from the Central Water Commission of the GoI (Thomas 2012: 45).

74. 'Mullaperiyar Damile Villalukal Bhoochalanam Moolam', *Mathrubhumi,* 21 December 2000, p. 1; 'Bhoochalanam Mullaperiyarine Badhichu, Vidagdha Samithi', *Malayala Manorama,* 21 December 2000, p. 9; 'Chorcha Adaykkanulla Tamil Nadinte Sramam Prathyakhathangalundakki', *Mathrubhumi,* 2 January 2001, p. 1; 'Mullaperiyar Dam Surakshithamallennu Niyama Sabha Samithi', *Malayala Manorama,* 2 January 2001, p. 9; K.T. Rajeev, 'Mullaperiyar, Kendram Vibhageeyatha Valarthunnu', *Deshabhimani,* 2 January 2001, p. 4.

75. 'Mullaperiyar, Tamil Nadu Villal Maraykkan Sramichathil Duroohatha', *Deshabhimani,* 21 December 2000, p. 1; 'Bhoochalanam Mullaperiyarine Badhichu', *Malayala Manorama.*

76. The committee had Kusala Rajendran, C.P. Rajendran, John Mathai, G. Sankar (all from the CESS), Sudhir Jain (IIT Kanpur), M.K. Parameswaran (retired KSEB member and the representative of Kerala in the Supreme Court commission), and Lalita Mickle (Chief Engineer [Design], Irrigation Department) as members. See 'Mullaperiyar, Suraksha Padikkan Samithi', *Deshabhimani,* 22 December 2000, p. 1; 'Mullaperiyar, Suraksha Parishodhanaykku Vidagdha Sangham', *Mathrubhumi,* 22 December 2000, p. 5; 'Mullaperiyar Anakkettinte Surakshaye Kurichu Padiykkan Vidagdha Samithi', *Madhyamam,* 22 December 2000, p. 5.

77. The strong influence of the state in national politics is a major reason for this. Regional parties like the DMK and the All India Anna Dravida Munnettra Kazhakam (AIDMK) have been a powerful presence in the national coalition politics. Kerala does not have such politically powerful regionalist parties.

78. 'Veendum Shakthamaya Bhoochalanam, Palayidathum Nasanashtam', *Malayala Manorama,* 8 January 2001, p. 1, lead article; Also see 'Bhoochalanam Veendum, Parakke Bheethi', *Mathrubhumi,* 8 January 2001, p. 1, lead article; 'Keralathil Veendum Bhoochalanam', *Deshabhimani,* 8 January 2001, p. 1, lead article; 'Veendum Bhoochalanam, Hridayakhatham Moolam Oru Maranam', *Madhyamam,* 8 January 2001, p. 1, lead article; 'Veendum Shakthamaya Bhoochalanam, Keralam Nadukkathil', *Kerala Kaumudi,* 8 January 2001, p. 1, lead article.

79. 'Veendum Shakthamaya Bhoochalanam', *Malayala Manorama.*

80. Peerumedu is closer to the Mullaperiyar dam.

81. The GSI team under the leadership of B.K. Shastri calculated the magnitude as 4.5 based on the data from the GSI's seismograph in Kulamavu. For them, the epicentre was between Erattupetta and Pala. According to the IMD, the intensity

was 4.8 and for the NGRI it was 4.9. Dr Rastogi also agreed with the GSI and the KSEB and was quoted by B.K. Sastri in the press conference. See 'Keralathil Veendum Bhoochalanam', *Deshabhimani*; 'Prabhava Kendram Palaykkum Erattupettaykkum Idayil', *Madhyamam*, 9 January 2001, p. 1; 'Bhoochalana Kendram Erattupettaykku Sameepam', *Mathrubhumi*, 8 January 2001, p. 5.

82. 'Keralathil Veendum Bhoochalanam', *Deshabhimani*; 'Mullaperiyar Damile Villal Valuthayi', *Deshabhimani*, 8 January 2001, p. 1; 'Mullaperiyar Villalukal Valuthayi', *Mathrubhumi*, 8 January 2001, p. 1; 'Mullaperiyaril Bhoochalanam Shaktham, Villalum Chorchayum Veendum', *Malayala Manorama*, 8 January 2001, p. 1; 'Mullaperiyaril Kooduthal Villalukal, Chorcha', *Madhyamam*, 8 January 2001, p. 1.

83. 'Mullaperiyaril Puthiya Anakkettu Nirmikkanam', *Mathrubhumi*, editorial, 9 January 2001, p. 4.

84. 'Mullaperiyar Damile Villal', *Deshabhimani*

85. 'Mullaperiyar, Vivekam Vediyaruthu', *Deshabhimani* 9 January 2001, p. 4, editorial. *Madhyamam* attempted to invoke the RIS argument to suggest that the Idukki region was highly vulnerable to earthquakes due to the presence of several dams, quoting scientific agencies such as the DMG and the CESS. See P.K. Prakash, 'Keralam Bhookampa Sadhyatha Meghala, Geology Padanam Njettikkunnathu', *Madhyamam*, 8 January 2001, p. 4.

86. 'Bhoochalanam Mullaperiyarilum Anubhavappettu: Report', *Kerala Kaumudi*, 23 January 2001, p. 12. Apparently it is the same report that was referred to by *Malayala Manorama* after two months. See A.B. Gijimon, 'Cheriya Bhoochalanavum Mullaperiyar Thakarkkum, Padana Report', *Malayala Manorama*, 23 March 2001, p. 1, lead article.

87. 'Bhoochalanam Mullaperiyarilum Anubhavappettu', *Kerala Kaumudi*.

88. V.S. Rajesh. 'Mullaperiyar Baby Dam Bhoochalana Sadhyatha Meghalayil, Jalanirappu Uyartharuthennu Vidagdha Samithi', *Madhyamam*, 19 February 2001, p. 1, lead article; 'Mullaperiyar Anakkettu Bhoochalana Meghalayil, Vidagdha Samithi', *Mathrubhumi*, 20 February 2001, p. 1.

89. 'Keralathinte Vaadathinu Melkai, Mullaperiyar Anthima Yogam Matti', *Malayala Manorama*, 24 January 2001, p. 11. *Malayala Manorama* alleged in another context, after the final report of the committee had been published, that the experiments performed by scientists from the CSMRS were 'scientifically invalid'. The newspaper even described how the 'pulse velocity test' was conducted by the team in order to point out the drawbacks of the test. See Jacob K. Philip, 'Anakkettu Apakadakaram, Dam Roopakalpana Chattangal Thanne Thelivu', *Malayala Manorama*, 11 February 2001, p. 1.

90. 'Keralathinte Vaadathinu Melkai', *Malayala Manorama*.

91. 'Mullapriyar, Jalanirappu Uyarthamennu Samithi', *Malayala Manorama*, 11 February 2001, p. 1, lead article; 'Jalanirappu 142 Adi Vare Uyartham, Vidagdha Samithi', *Madhyamam*, 11 February 2001, p. 1, lead article.

92. 'Viswasangalkku Villal, Keralavum Bhoochalana Meghala', *Malayala Manorama*, 8 January 2001, p. 14; 'Bhoochalanam, Jagratha Palikkanam', *Malayala Manorama*, 9 January 2001, p. 8, editorial; 'Dakshinendia Ake Bhoochalana Bheeshaniyil', *Kerala Kaumudi*, 8 January 2001, p. 1; 'Oru Masathinidayil Randam Thavana', *Kerala Kaumudi*, 08 January 2001, p. 1.

93. 'Keralam Jagratha Pularthanam: Dr. Srivastava', *Mathrubhumi*, 8 January 2001, p. 1.

94. 'Ithu Asadharana Anubhavam', *Mathrubhumi*, 8 January 2001, p. 8.

95. 'Bhoochalanangal Avarthikkan Sadhyatha, Rastogi', *Mathrubhumi*, 8 January 2001, p. 8.

96. 'Bhoochalanangal Ore Disayil', *Deshabhimani*, 8 January 2006, p. 1.

97. 'Keralam Bhookampa Sadhyatha', *Madhyamam*.

98. 'Oru Masathinidayil Randam', *Kerala Kaumudi*.

99. 'Bhoochalanam, Urul Pottal Sadhyatha Eri: Sastrajjnar', *Deshabhimani*, 10 January 2001, p. 6; 'Sakthamaya Bhoochalanathinu Sadhyathayilla', *Kerala Kaumudi*, 12 January 2001, p. 12; 'Bhoomi Veendum Kulungumpol Tentukal Abhaya Kendram', *Mathrubhumi*, 24 January 2001, p. 1.

100. R. Prakasan, Chirayinkeezhu, 'Sambhranthi Parathunna Bhoochalanangal', *Kerala Kaumudi*, 14 January 2001, p. 4, letter to the editor.

101. Dr M. Manikantan Nair, Thiruvananthapuram, 'Bhoochalanathil Kulungunna Sastrajjnar', *Mathrubhumi*, 13 January 2001, p. 4, letter to the editor.

102. Similar attacks were also made directly by the newspapers. For instance, *Deshabhimani* in a report pointed out that scientists at the CESS were capable of undertaking only a preliminary analysis and that was why there was great discrepancy in their explanations. See 'Bhoochalanam, Urul Pottal', *Deshabhimani*.

103. The reader argued that tectonic activity was intense under mountains and hence the presence of the Western Ghats was important.

104. 'Bhoochalanam, Urul Pottal', *Deshabhimani*.

105. See, for instance, Thayyullathil Abbas, 'Bhoochalanam: Munkaruthal Venam', *Madhyamam*, 19 January 2001, p. 4, letter to the editor.

106. Sasidharan Mangathil, 'Bhookampam Pravachikkam', *Mathrubhumi*, 8 January 2001, p. 4, editorial page article.

107. Sajan, 'Bhoochalana Pravachanam, Chinese Reethikku Prasakthi Erunnu', *Deshabhimani*, 16 January 2001, p. 5.

108. In the post-Gujarat earthquake phase of deliberation too, several such cases were reported. *Madhyamam* pointed out the successful scientific prediction of the Blue Mountain Lake earthquake in New York (1973) and the Haicheng earthquake in China (4 February 1975). See 'Bhookampa Pravachanam: Vidagdharkku Bhinnabhiprayam', *Madhyamam*, 29 January 2001, p. 5. A reader explained how he witnessed an earthquake being forecasted with great precision in Japan. See M.J. Mathew, 'Pravachicha Bhoochalanam

Nerittarinjappol', *Malayala Manorama*, 31 January 2001, p. 8, letter to the editor. In response to this letter, another reader pointed out that Japan benefits from traditional wisdom concerning earthquake forecasts, and that by observing the behaviour of a species of fish called the 'living seismograph', they could forecast tremors. See K. Mano Mohan, 'Sahayikkunnathu Meen!', *Malayala Manorama*, 9 February 2001, p. 8, letter to the editor. Scientific studies were cited which suggested that changes in the behavioural patterns of rodents, reptiles, birds, and cattle could be used to anticipate earthquakes. 'Bhookampam Ethum Mumpe Jeevikal Ariyunnu', *Mathrubhumi*, 28 January 2007, p. 15; 'Bhookampa Soochana Nalkan Oru Pakshi', *Kerala Kaumudi*, 28 January 2001, p. 4; 'Mrigangal Karanju, Aanakal Irunnu', *Kerala Kaumudi*, 28 January 2001, p. 4. The possibility of developing a reliable prediction mechanism based on the abnormal behaviour of these creatures was also discussed alongside the possibility for prediction based on radon gas emission from the earth. 'Bhookampam Ethum Mumpe Jeevikal Ariyunnu', *Mathrubhumi*, 28 January 2007, p. 15; 'Bhookampathinte Novarinjathu Charles Richter', *Malayala Manorama*, 28 January 2001, p. 4.

109. See the large number of reports in the Malayalam newspapers between 9 January 2001 and 5 February 2001.

110. 'El Salvadoril Bhookampam: 234 Per Kollappettu', *Deshabhimani*, 15 January 2001, p. 1; 'Bhookampam, Madhya Americayil Maranam 138 Ayi', *Mathrubhumi*, 15 January 2001, p. 8; 'Bhookampam: Maranam 400 Kavinju, Ayirangale Kananilla', *Malayala Manorama*, 16 January 2001, p. 16; 'Bhookampathil Maranam 600', *Malayala Manorama*, 17 January 2001, p. 18; 'El Salvadoril Bhookampam: 138 Maranam', *Madhyamam*, 15 January 2001, p. 5; 'El Salvadoril Bhookampam: Maranam 381 Ayi', *Madhyamam*, 16 January 2001, p. 8; 'El Salvadorilum Guatemalayilum Bhookampam: 234 Maranam', *Kerala Kaumudi*, 15 January 2001, p. 1.

111. The same reporting strategy has been repeated on other occasions too to create a greater impact. See Chapter 7 for more discussion on this.

112. 'Bhookampathinte Theevratha: Karyamariyathe Thettidharana', *Malayala Manorama*, 28 January 2001, p. 1; 'Bhookampa Vyapthiyekurichu Tharkkam', *Deshabhimani*, 28 January 2001, p. 6; 'Gujarathil Rekhappeduthiyathu 7.9 point', *Madhyamam*, 28 January 2001, p. 5.

113. 'Gujarathinte Vedana: Oro Bharateeyanteyum Dukham', *Mathrubhumi*, 28 January 2001, p. 4, editorial.

114. 'Rajyathe Nadukkiya Bhoochalanam', *Malayala Manorama*, 28 January 2001, p. 4, editorial; 'Keralavum Gujarathumayi Samanathakalilla', *Mathrubhumi*, 28 January 2001, p. 5; Also see Sasidharan Mangathil, 'Kutch Van Bhookampa Meghalayil', *Mathrubhumi*, 28 January 2001, p. 4, editorial page article.

158   Contested Knowledge

115. Mangathil, 'Kutch Van Bhookampa', *Mathrubhumi*; P.K. Prakash, 'Bhookampam: Adiyanthira Suraksha Nadapadi Anivaryam,' *Madhyamam*, 28 January 2001, p. 4, editorial page article. News reports on the scientific report submitted by two scientists from the CESS to the Department of Science and Technology (DST) which suggested the possibility of a major earthquake in the Rann of Kutch and Bhuj areas of Gujarat also appeared in the press. See 'Malayali Sastrajjnanmar Munnariyippu Nalkiyirunnu', *Kerala Kaumudi*, 28 January 2001, p. 4; 'Malayali Sastrajjnanmarude Munnariyippu Avaganichu', *Malayala Manorama*, 28 January 2001, p. 9.

116. 'Keralavum Gujarathumayi Samanathakalilla', *Mathrubhumi*; 'Gujarathinte Vedana: Oro', *Mathrubhumi*. Also see Mangathil, 'Kutch Van Bhookampa', *Mathrubhumi*.

117. 'Keraleeyar Bhayappedendathilennu Sastrajjnar', *Kerala Kaumudi*, 28 January 2001, p. 1; 'Keralavum Gujarathumayi Samanathakalilla', *Mathrubhumi*, 28 January 2001, p. 5; 'Keralathil Asanka Venda', *Deshabhimani*, 28 January 2001, p. 7; 'Bhookampam: Samsthanathu Asanka', *Madhyamam*, 28 January 2001, p. 7.

118. For instance, see A.K. Manoj, 'Bhookampa Sadhyathakalum Munkaruthalukalum', *Mathrubhumi*, 31 January 2001, p. 4; K.N. Krishnakumar, 'Bhookampam: Pravachanatheethamaya Prathibhasam', *Deshabhimani*, 31 January 2001, p. 4, editorial page article; 'Bhoochalanam: Keralathinu Asanka Venda: GSI Director General', *Mathrubhumi*, 31 January 2001, p. 5; 'Bhookampam: Gujarathumayulla Tharathamyam Arthasoonyam', *Malayala Manorama*, 31 January 2001, p. 4.

119. Harishankar T.S., 'Bhookampa Pravachana Sadhyatha Ezhuthi Thallaruthu', *Mathrubhumi*, 5 February 2001, p. 4, letter to the editor; T.S. Harishankar, 'Bhookampathe Kurichu Munnariyippu Nalkan Kazhiyanam', *Kerala Kaumudi*, 6 January 2001, p. 4, letter to the editor.

120. 'Vyaja Pracharanam: Keralathil Paribhranthi', *Deshabhimani*, 30 January 2001, p. 1; 'Keralathil Bhoochalana Sadhyatha Illa', *Deshabhimani*, 30 January 2001, p. 1; 'Engum Paribhranthi', *Mathrubhumi*, 30 January 2001, p. 1; 'Jillayil Vyaja Sandesham Bhookampa Bheethi Parathi', *Mathrubhumi*, 30 January 2001, p. 2; 'Abhyoohangal Pracharippikkaruthu: Collector', *Mathrubhumi*, 30 January 2001, p. 2; 'Keralathil Bhookampathinu Sadhyathayilla', *Mathrubhumi*, 30 January 2001, p. 5; 'Bhoochalanam: Keralathinu Asanka', *Mathrubhumi*; 'Bhookampam: Kimvadanthi Engum Paribhranthi Parathi', *Malayala Manorama*, 30 January 2001, p. 2; 'Bhoochalanam: Nuna Sandeshangalil Keralam Bhayannu', *Malayala Manorama*, 30 January 2001, p. 9; 'Vyaja Bhookampa Pracharanangalil Dakshina Keralam Kulungi', *Madhyamam*, 30 January 2001, p. 1; 'Abhyoohangal Pracharippikkaruthu', *Madhyamam*, 30 January 2001, p. 5; 'Bhookampam Pravachikkan Kazhiyilla: Sastrajjnar', *Kerala Kaumudi*, 30 January 2001, p. 1; Bhookampa Nuna Keralathil Paribhranthi Parathi', *Kerala Kaumudi*,

30 January 2001, p. 1; 'Iniyoru Irupathu Varsham Kazhinjalum Bhookampam Pravachiykkanavilla', *Kerala Kaumudi*, 31 January 2001, p. 10.

121. See 'Keralathil Bhoochalana Sadhyatha Tallikkalayanavilla', *Madhyamam*, 31 January 2001, p. 7.

122. 'Keralathil Bhoochalana Sadhyatha', *Madhyamam*.

123. 'Nadee Samrakshanavum Vanavalkaranavum Bhookampa Sadhyatha Kuraykkum: Qutabdin', *Mathrubhumi*, 3 February 2001, p. 9.

124. Dr S. Mukherjee was an Associate Professor in the School of Environmental Sciences of the Jawaharlal Nehru University. Later he published a scientific paper on his theory of the influence of earth-directed coronal mass ejection and its effect on the planet's magnetic field. See Mukherjee (2008).

125. See Anish M. Ali, 'Bhookampangal Undavunnathu', *Madhyamam*, 5 February 2001, p. 4, editorial page article.

126. 'Bhoochalanam Iniyum Undakumennu Munnariyippu', *Malayala Manorama*, 8 January 2001, p. 1.

127. 'Bhookampam Munkootti Ariyam: Sastrajjnan', *Malayala Manorama*, 3 February 2001, p. 14.

128. 'Bhookampam Pravachikkam: P.N. Nair', *Madhyamam*, 4 February 2001, p. 12.

129. The name of the theory might be an allusion to the 'Raman Effect' discovered by Dr C.V. Raman, for which he won the Nobel Prize for Physics in 1930.

130. K.P. Devdas, 'Bhookampavum Vavum', *Mathrubhumi*, 3 February 2001, p. 4, letter to the editor.

131. P. Sajeev, Madathara, 'Durantha Karanam Anupareekshanam', *Deshabhimani*, 2 February 2001, letter to the editor.

132. Another letter which expressed a similar concern was from Shumin S. Babu, 'Anupareekshanavum Bhookampavum Thammil Bandhamundu', *Deshabhimani*, 14 February 2001, p. 4, letter to the editor.

133. 'Durantham Vannathu Magha Masathile Krishnapaksha Chathurthiyil', *Mathrubhumi*, 29 January 2007, p. 14.

134. S. Joykutty, Edamanni, 'Cheythathu Sariyo?', *Malayala Manorama*, 10 February 2001, p. 8, letter to the editor.

135. Counterarguments of some scientists such as Dr S.K. Srivastava of the IMD were published. He upheld the uniqueness of earth science by distancing it from non-sciences like astrology and underscored that earthquakes cannot be predicted. See 'Bhookampam Pravachikkanavilla: S.K. Srivastava', *Malayala Manorama*, 29 January 2001, p. 3.

136. This does not imply that astrology is not a strong presence in the public domain. Astrology is rigorously practised in Kerala, and many of the television channels have regular programmes related to astrology. My argument is that

astrology did not enjoy the status of a knowledge tradition with capabilities of
forecasting disasters, in the scientific public sphere.

137. See the concept of 'civic epistemological frames' that further helps us
understand the gate-keeping mechanism of the scientific public sphere, discussed in the next chapter.

138. 'Thirayilakkam Bhookampamalla', *Madhyamam*, 19 February 2001,
p. 2. Dr P.K. Rajendran Nair, 'Keralathile Cheruchalanangal Pradeshika Prathibhasam', *Malayala Manorama*, 19 February 2001, p. 8, editorial page article;
'Bhoochalanamalla, Bhouma Kampanamennu CESS Adhikrithar', *Madhyamam*,
13 February 2001, p. 1; 'Bhoochalanam: Asanka Vendennu Vidagdhar', *Madhyamam*, 18 February 2001, p. 2; 'Kinarukalile Thirayilakkam Thudarum, Paribhranthi Venda', *Kerala Kaumudi*, 16 February 2001, p. 2; 'Keralathil Bhoochalana
Sadhyatha Kuravu', *Deshabhimani*, 21 March 2001, p. 7.

139. 'Kinarukalil Vellam Alakaluyarthi Pongunnu: Kangarappadiyil
Janangal Bheethiyil', *Malayala Manorama*, 31 January 2001, p. 3.

140. 'Idamalayaril Mala Pukayunnu: Rannykkaduthu Bhoomi Thazhunnu',
Malayala Manorama, 19 July 2001, p. 9; 'Pukayunna Malaykkadutha Kinattile
Vellam Vatti: Vaikkathu Kinar Idinju', *Malayala Manorama*, 20 July 2001,
p. 1; 'Malapukayal: Vidagdha Padanam Avasyamennu Vidagdha Sangham',
Madhyamam, 22 July 2001, p. 5; 'Idinjimalayile Pukapadalam Manjanennu
Varuthan Shramam', *Deshabhimani*, 24 July 2001, p. 2; 'Idinjimalayile Puka
Moodalmanjennu "CESS" Sastrajjnar', *Madhyamam*, 25 July 2001, p. 5.

141. 'Idinjimalayile Puka Moodalmanjennu', *Madhyamam*.

142. 'Idinjimalayile Puka Moodalmanjennu', *Madhyamam*.

143. 'Idinjimalayile Puka Moodalmanjennu', *Madhyamam*.

144. See 'Idinjimalayile Puka Moodalmanjennu', *Madhyamam*.

145. 'Kinarukalil Jalamuyarunnathu Bhoochalanathinte Thudar Prathibhasam', *Malayala Manorama*, 1 February 2001, p. 9.

146. See the extensive reporting from different parts of the state in the latter
half of February 2001.

147. See the reports on earthquakes in China, Japan, Vietnam, Indonesia,
Malaysia, El Salvador, Taiwan, Central Asia, the USA, and Canada, which
appeared in the regional press during the same period. The eruption of the
Merapi volcano (Java, Indonesia) also was reported.

148. 'Bhoogarbha Chalanangal Thudarunnathu Apathsoochana Thanne',
Mathrubhumi, 15 February 2001, p. 4, editorial.

149. Valiyavila Anil, Vattiyoorkavu, 'Bhoomiyude Munnariyippu Avagani-
kkaruthu', *Mathrubhumi*, 16 February 2001, p. 4, letter to the editor.

150. Sasidharan Mangathil had his postgraduate training in geology from
the University of Calicut, Kerala, and later joined a research project in the
CESS. He had also had field experience in petroleum excavations while working

as a technician in an oil excavation company (Mumbai Eastern Circuit) that collaborated with the Oil and Natural Gas Corporation (ONGC) of the GoI. Slowly he shifted to journalism and later joined *Mathrubhumi*. As a journalist he specializes on environmental issues as well as on landslides and related geological issues, and has won many awards for environmental journalism. The series with the title 'Keralam Kulungumpol' ('When Kerala Shudders') was published from 17 February 2001 to 23 February 2001, with six essays in the editorial page. The series discussed the history of earthquakes, the geological structure of Kerala, RIS thesis, and so on. It also pointed out the sorry state of geological research in the state and made suggestions for effective research, including the preparation of a microzonation map. He has also published a book on the Mullaperiyar water dispute (Mangathil 2008).

151. Sasidharan Mangathil, 'Keralam Kulungumpol–1', *Mathrubhumi*, 17 February 2001, p. 4.

152. The *Mathrubhumi* also published several readers' letters which shared similar concerns, apparently to demonstrate the growing public dissent. See, for example, Dr R. Gopinathan, 'Swayam Srishtikkunna Duranthangal', *Mathrubhumi*, 24 February 2001, p. 4, letter to the editor; Mulloor Surendran, Vizhinjam, 'Bhoomiyude Hridayam Vingumpol', *Mathrubhumi*, 28 February 2001, p. 4, letter to the editor; Dr A. Manikandan Nair, Thiruvananthapuram, 'Bhoumasastrajjnar Bhoomiyude "Vakthakkal" Aakukayo?', *Mathrubhumi*, 24 February 2001, p. 4, letter to the editor. The last letter strongly criticized scientists for their irresponsible statements that were not based on 'solid research'. The reader also urged them to take lay-expertise seriously in order to reach sound conclusions.

153. However, as will be discussed later in the chapter, there were serious attempts to break the mould of the deficit model by some experts who allied with the media to address the public concerns.

154. However, they are an extremely marginal presence in mainstream journalism in Kerala. Their number may grow in the future because of the increasing frequency of risk controversies and science reporting in general in mass media. For more on the subject, see Chapter 7.

155. This was also alternatively referred to as *Kinar Thazhal* or *Kinar Idiyal*.

156. 'Kinarukal Aprathyakshamakumpol', *Mathrubhumi*, 19 June 2001, p. 4, editorial.

157. 'Samsthanathu 123 Kinarukal Thazhnnu', *Malayala Manorama*, 13 July 2001, p. 11.

158. The phenomenon of coloured rain that had begun appearing around this period developed into a major scientific controversy in its own right and hence it will be discussed separately in the next chapter. For a complete list of the phenomena, see Appendix.

159. The case of the fuming of a hill called Idinjimala in Irattayar (Idukki district) that we have discussed earlier in the chapter had occurred during this time.

160. 'Bhoomiyude Apabhramsam Moolamalla: Vidagdhar', *Deshabhimani*, 20 June 2001, p. 7; 'Kinarukal Aprathyakshamayathu Bhoojalamardham Moolam: Vidagdha Samithi', *Mathrubhumi*, 21 June 2001, p. 1; Kinarukal Thakarnnathu Bhoojalamardham Moolamennu Vidagdhar', *Malayala Manorama*, 21 June 2001, p. 9.

161. 'Kinarukal Thakarnnathu Bhoojalamardham', *Malayala Manorama*.

162. 'Kinar Aprathyakshamakal Bhoomiyude Puramthodile Chalanam Moolamennu Padanam', *Mathrubhumi*, 7 July 2001, p. 1.

163. The research team was led by the Senior Hydro Geologist of the department, Dr K.V. Mohanan, and the other geologists in the team were Anto Francis, K.K. Sajeevan, and K.M. Ashraf.

164. See 'Kinar Thazhchaykku Bhookampavumayi Bandham: CESS', *Deshabhimani*, 5 August 2001, p. 7. The *Deshabhimani* has published a report that showed the contrast between the two scientific studies. See 'Kinar Thazhunnathu Bhookampam Moolamallennu Padana Report', *Deshabhimani*, 9 July 2001, p. 7.

165. 'Kinar Thazhunnathu Bhookampam', *Deshabhimani*.

166. 'Kinar Thazhchaykku Bhoochalanavumayi Bandhamilla: CESS', *Deshabhimani*, 24 July 2001, p.12.

167. 'Kinar Thazhchaykku Bhoochalanavumayi', *Deshabhimani*.

168. 'Kinar Thakarcha Bhoochalanathinte Prathikaranamalla', *Mathrubhumi*, 22 July 2001, p. 5. Nonetheless, he acknowledged the sorry state of geological research in Kerala and emphasized the need for comprehensive research on the phenomena.

169. 'Kinar Thakarcha Bhoogarbha Jalamardham Moolam: Report', *Malayala Manorama*, 20 July 2001, p. 1.

170. 'Kinarukalum Kulangalum Aprathyakshamakumpol', *Malayala Manorama*, 20 July 2001, p. 8, editorial.

171. 'Bhoogarbha Prathibhasangalil Vidagdhanweshanam Avasyam', *Mathrubhumi*, 13 July 2001, p. 4, editorial.

172. 'Aswabhavika Prathibhasangal Bheethi Padarthunnu', *Deshabhimani*, 29 July 2001, p. 10.

173. 'Idiyunna Kinarukal Nalkunna Soochana', *Madhyamam*, 13 July 2001, p. 4, editorial.

174. See the letters to the editor that appeared in the month of July in the newspapers.

175. The competition between *Malayala Manorama* and *Mathrubhumi* for attracting more readers and thereby attaining more advertisement revenue has

been observed by Jeffrey (2000). The cut-throat competition was also due to the fact that they together held around 87 per cent of the total circulation of newspapers in Malayalam. Reporting of exclusive stories is integral to the competition between the newspapers.

176. See the series of essays published under the title 'Keralam Kulungumpol', written by Sasidharan Mangathil from 17 February 2001 to 23 February 2001.

177. 'Keralathil Bhoovillalukal Sajeevamakunnu: Sastrajjnarkku Asanka', *Mathrubhumi*, 27 July 2001, p. 1.

178. 'Keralathil Bhoovillalukal Sajeevamakunnu', *Mathrubhumi*.

179. 'Keralathil Bhoovillalukal Sajeevamakunnu', *Mathrubhumi*. It is interesting to note that the same scientists were quoted later by some other newspapers as arguing that the phenomena would not cause any havoc, since it simply indicated the tectonic adjustments of the earth. See 'Apoorva Prathibhasangalil Asanka Venda: Sastrajjnar', *Deshabhimani*, 8 August 2001, p. 8; 'Prakrithi Prathibhasangalil Paribhranthi Vendennu Sastrajjnar', *Madhyamam*, 8 August 2001, p. 5; 'Kinar Idichilum Varna Mazhayum: Asanka Vendennu Vidagdhar', *Malayala Manorama*, 9 August 2001, p. 2.

180. 'Bhoogarbha Chalanangal: Karuthal Nadapadikal Venam', *Mathrubhumi*, 28 July 2001, p. 4, editorial.

181. P. Balakrishnan, 'Kinaridichil Bhookampathinte Munnodiyavam', *Mathrubhumi*, 30 July 2001, p. 1.

182. See K.V. Mohanan, 'Kinarukal Thakarunnathu Enthukondu?' *Mathrubhumi Varanthappathippu*, 30 September 2001, p. ii.

183. 'Kinar Thazhchaykku Bhookampavumayi Bandham', *Deshabhimani*, 5 August 2001, p. 7; 'Kinar Idiyalinu Bhookampavumayi Bandham', *Madhyamam*, 9 September 2001, p. 12.

184. See the wide range of reports and articles in the regional press presenting the fault line theory as a strong argument against the expert opinion of the research team constituted under the directive of the CM of Kerala. However, this development did not prevent alternative scientific propositions regarding the phenomenon. For instance, *Deshabhimani* published an article in which the authors suggested the tectonic activities in the 'Ring of Fire' region of the Arabian-Indo-Australian Plate as causing instability in the 'continental shelf'. The vibrations from this zone, according to them, were responsible for the water oscillations and the well collapses in Kerala, a region which was already vulnerable due to continuing earthquakes and the presence of active fault lines. See S. Suresh Kumar and Dr A.M. Nair, 'Kinarukalum Kulangalum Enthukondu Aprathyakshamakunnu?' *Deshabhimani*, 1 August 2001, p. 4, editorial page article. The same scientists, in a letter to the editor in another newspaper, explained the red rain phenomenon in terms of the Ring of Fire

theory. See Chapter 6. N.J.K. Nair, former Director of the resource analysis wing of the CESS, propounded a hypothesis regarding the weakening of aquifers, an underground layer of permeable rock that stored ground water. Because of deforestation, he argued, the aquifers failed to get refilled, and eventually weakened. The well collapses as well as the water oscillation were explained as a trivial, harmless, and localized phenomenon on the basis of this hypothesis. See Sajan Maranadu, 'Kinarukal Kanathavunnathu', *Varadya Madhyamam*, 16 September 2001, p. 3.

185. The CESS Director, however, kept on opposing the fault line theory and reinstated the groundwater pressure theory. See 'Kinar Kanathavunnathu Bhookampa Soochanayalla: CESS director', *Malayala Manorama*, 5 August 2001, p. 9.

186. See 'Kinar Kanathakal: Padanam Nadathan "Manorama" Niyogicha Sasthra Sangham Ethi', *Malayala Manorama*, 12 August 2001, p. 1.

187. Dr Negi was Emeritus Scientist at the NGRI, Hyderabad, and an awardee of the prestigious Shanti Swarup Bhatnagar prize (1980) of the Council of Scientific and Industrial Research (CSIR). Dr Bapat is a renowned seismologist and formerly the Chief Research Officer of the Earthquake Engineering Research Division of the Central Water and Power Research Station at Pune.

188. The *Malayala Manorama* reported the investigations and other activities of the expert team on a daily basis. The research team also met the CM, A.K. Antony. See the reports in *Malayala Manorama* from 12 to 17 August 2001.

189. B. Ajith Babu, 'Kinar Thazhal: Bhoogarbha Viseshangalude Puram Kazhcha', *Malayala Manorama*, 17 August 2001, p. 16, editorial page article.

190. Babu, 'Kinar Thazhal: Bhoogarbha', *Malayala Manorama*.

191. In the earlier dispute over the epicentre and magnitude of the earthquake too, a similar arbitration had been carried out, as discussed in the first part of the chapter.

192. Babu, 'Kinar Thazhal: Bhoogarbha', *Malayala Manorama*.

193. 'Kinar Thazhunnathinepatti Kooduthal Padanam Venam', *Malayala Manorama*, 21 August 2001, p. 9, editorial.

194. 'Kinarukal Idiyunnathu Bhramsana Meghalakal Kendreekarich', *Mathrubhumi*, 23 August 2001, p. 1.

195. 'Sastrajjna Sanghathe Niyogikkanam', *Mathrubhumi*, 24 August 2001, p. 4, editorial.

196. For incidents of well collapse and related phenomena reported a few years later, see 'Kayaril Thookkiyitta Kinar', *Malayala Manorama*, 16 February 2006, p. 4 (New Delhi edition); 'Kinaril Thirayilakkam: Veettukar Paribhrantharayi', *Mathrubhumi*, 31 May 2007, p. 8 (New Delhi edition). The Government of Kerala commissioned the Kerala State Remote Sensing and Environment Centre (KSRSEC), an institute under the Kerala State Land

Use Board, during 2005–7 to conduct a detailed study of the phenomena. See 'Kinaridichil: Karanam Thedi Visadamaya Padanam Thudangi', *Malayala Manorama*, 12 February 2006, p. 9 (New Delhi edition). I was unable to access the findings of the study.

6 Border Crossings
The Coloured Rain Controversy

So far we have seen two key aspects of public engagement with science. In the clinical trials controversy, the need for public regulation of research was asserted against the backdrop of the changing configuration of science in the neoliberal context. The earthquake controversy that proliferated around contradictory scientific explanations eroded the public trust in experts. Although supra-scientific authorities were created to arrest the erosion of scientific credibility, an epistemological conflict that pitched the experts and the scientific-citizen public on oppositional sides resulted. Mass media played an axiomatic role in shaping both the controversies. In the former case, the regional press was divided in their allegiance to the two groups of researchers at loggerheads with each other. In the latter, newspapers aligned themselves with the public sentiments and perceptions of risk, challenging experts' assurances, and mobilizing scientific resources to facilitate new interpretations that addressed the public's anxieties and risk perceptions.

In this chapter, we will be exploring the medialization of science further through the coloured rain controversy which was linked to the latter case of tremors and well collapses. The initial explanation for coloured rains in the region was a meteor explosion, but this interpretation was challenged afterwards on various grounds. Wide-ranging scientific and quasi-scientific explanations were offered, especially when a group of biologists presented evidence for the existence of large quantities of biological cells in the coloured rainwater samples. The endeavour to scientifically explain the phenomenon was a long drawn out process of negotiation with a wide array of actors. Media turned out to be the main channel of communication for scientists, who belonged to different disciplinary and institutional backgrounds, during the manifestation of the unnatural phenomenon. In their endeavour to scientifically analyse the causes of the phenomenon, institutional collaborations were formed and broken. Public credibility of the Centre for Earth Science Studies (CESS), Thiruvananthapuram, that had already been in the decline during the earthquake controversy got further detriorated with the coloured rain controversy. Public deliberation about the scientific causes of the coloured rain also revealed the interpretive latitude surrounding scientific explanations and the strategies that were formulated by diverse actors to challenge dominant scientific claims. This has continuously forced the established boundaries between disciplines, subcultures, and networks of science open for negotiation. The controversy witnessed the emergence of a small group of researchers skilfully highlighting their quasi-scientific theory in public and forming coalitions with researchers in Europe. These 'border crossings' facilitated by medialization of science is the focus of this chapter.

Meteor as the Villain

When the scientific public sphere was debating the question of tremors and well collapses, a new phenomenon unexpectedly appeared. It was first reported from Changanacherry, a small town in the Kottayam district. The inhabitants of Morkulangara, a locality within the town, experienced a peculiar 'red rain' (*chuvanna mazha* or *chuvappu mazha*) on Wednesday morning, 25 July 2001.[1] The event did not receive much attention in the regional press the next day, and was generally reported only in the local editions of the dailies.[2] However, the incident moved to the front page after scientists from the CESS visited the site for investigations the day after the incident. By that time, similar incidents of rain in different colours had occurred in other parts of the state too and this contributed to its increased news worthiness.[3]

Dr V. Sasi Kumar and Dr S. Sampath, scientists from the CESS,[4] arrived at Changanacherry and collected the rainwater samples. The main question they were supposed to answer was the relationship between this new phenomenon and the tremors, the well collapses, and the 'tunnel/well formations' in the region. They opined that it was a rarest of the rare phenomenon, which had never before been reported in India, and refuted any possible linkage of red rain with the other unusual events that had recently been observed in Kerala.[5] The presence of dust particles, chemicals, or microorganisms in the atmosphere was seen to be the cause of the red rain.[6] They pointed out that similar incidents had been reported in the past from England and Italy. It is interesting to note that the scientists collected background information about the phenomenon from the internet, indicating that the red rain was a completely new phenomenon even for the experts.[7]

In the days that followed, newspapers trailed the phenomenon more closely. The media also kept a close tab on the investigations and explanations of the CESS scientists. Red rains became quite a frequent occurrence in Kerala within a day or two of the first incident.[8] Along with the news reports on coloured rains, analytical reports also appeared in the regional press,[9] revealing the crucial role of media in constructing the phenomenon as one of great public importance. Newspapers reported that the main thrust of scientific research undertaken by the expert team from the CESS was to elucidate the source of origin of the red colour in the rains.[10] The news reports also informed the public that the red sediment found in rainwater samples would be extracted and analysed by experts.[11] *Deshabhimani* quoted Dr Sasi Kumar as suggesting the presence of suspended particulate matter in air as the causative agent, but also cautioning that more research was needed to confirm the origin of the phenomenon.[12]

Other scientific explanations of the phenomenon began to appear at this juncture. Scientists from different disciplinary backgrounds participated in the deliberations on the causative factors and, interestingly, the scientific public sphere turned into the prime site for such interdisciplinary exchanges. Dr B. Mukhopadhyay, a pollution expert from the India Meteorological Department (IMD), suggested the presence of industrial pollutants such as nitrogen dioxide and red coloured soil particles in the atmosphere as the causative agents of this short-term phenomenon.[13] He pointed out that a similar incident had been reported in 1991 from Ernakulam.[14] Another environmental scientist argued that it was a biological phenomenon, and, therefore, suggested that a biological assay be undertaken rather than a chemical analysis.[15]

For him the red colour appeared when the litmus produced by certain lichens reacted with acid molecules in contaminated rain. With the regional press acting as mediators of scientific views that were otherwise communicated among experts in scientific conferences and interdisciplinary journals, a more public mode of research communication opened up, revealing the internal polysemy of the knowledge production process to the public. This also indicated the non-availability of any communication channels internal to science for the exchange of ideas and opinions in such contexts of manifestation of risks. The sudden appearance of a new natural phenomenon and the risks it raised urged the scientific community to express their ideas and access others' interpretations through the regional press (cf. Lewenstein 1995a). The scientific public sphere, thus, became a site for knowledge production.

The inter-specialist scientific communication (Cloitre and Shinn 1985) with the help of mass media invited more non-scientist actors to participate in the exchange in the following days. The newspapers also actively participated in the debate in their capacity as actors, often highlighting the internal ambivalences of the scientific community and their inability to provide a consensual interpretation of the phenomenon.[16] By situating the 'coloured rain' within the wider landscape of negotiation regarding the vulnerability of the geological structure of the region, *Deshabhimani* gestured towards the tacit connections between earthquakes, well collapses, and coloured rains.[17] News reports in *Deshabhimani* and *Mathrubhumi* challenged the assumption that it was an environmental issue rather than a geological one, pointing out the ambiguity regarding the origin of the suspended matter.[18] In this context Dr M. Baba, the director of the CESS, argued that the red rain was a risk-free phenomenon for a number of reasons.[19] He refuted both the acid rain and the microbial hypotheses and hinted at the possible presence of residual particulate matter from a meteor in his attempt to defend the researchers from the institute.[20]

The preliminary report of the researchers from the CESS on the red rain in Changanacherry was released at this point of time. The report concluded that it was caused by the dust from a passing meteor.[21] The CESS scientists explained that about 1,000 kilograms of ash from the burning meteor diffused in the atmosphere and was washed down by the rain that followed.[22] The red colour of the rain was due to the presence of a large amount of meteor ash in the rainwater, according to them.

The researchers explained that the unusual thunderclap and the flash of light reported at 5:30 a.m. on the day of the red rain were caused by the meteor explosion above Madhumoola Junction in Changanacherry. Based on meteorological reports, the CESS scientists pointed out that there was no likelihood of a thundershower at that time in the locality and suggested this as substantiating their argument about the meteor explosion.[23] The explanation triggered off a big debate in the scientific public sphere and the persistence of coloured rains with greater frequency all over Kerala catalysed it.

The CESS Scientists as Villains: Emergence of Counterarguments

While all the other newspapers presented the findings of the CESS scientists in a plain, matter-of-fact manner, *Mathrubhumi*'s reporting was in an entirely different style. The report on the first page of the newspaper presented the meteor hypothesis in opposition to the argument of the scientists from the School of Applied Life Sciences (SALS) of Mahatma Gandhi University, Kottayam, generating a polemic debate.[24] The Director of the SALS, Dr K.C. John,[25] reasserted the hypothesis that biological cells in the rain had coloured the rainwater. This was based on the analysis of the coloured rainwater under an electron microscope. However, the sample of coloured rain was collected not from Changanacherry but Valanchuzhi, a place near Pathanamthitta, where the institute was located.[26] The biologists identified three different types of cells in the sample—small and translucent as well as yellow and brick red coloured ones. They also noticed cell division in some of the cells. The SALS scientists concluded that the cells were either of some algal or fungal origin or were pollen grains.[27] The intervention of the life scientists from the university department had a negative impact on the validity of the argument of the meteorologists from the CESS. The epistemic authority of the meteorologists was challenged by the biologists, and the presence of a higher order scientific instrument like the electron microscope further helped them legitimize the biological cell argument. Moreover, the CESS had already been suffering from a loss of credibility in the scientific public sphere because of their interventions in the earthquake controversy. This might have aggravated the epistemological vulnerability of the CESS scientists, further diminishing public trust in them. The logically coherent argument of the CESS scientists enjoyed lesser public appreciation than the biologists' explanation that was actually inconclusive since they could not identify the species to which the cells/spores belonged.

The controversy gathered momentum as more scientists and scientific institutions joined the debate in the following days, and a new alliance was formed amongst the biologists against the meteorologists and earth scientists.[28] Reports on the confirmation of the SALS's argument by researchers from the Central Plantation Crop Research Institute (CPCRI), Kasaragod, appeared in the media.[29] Advanced collaborative research on the coloured rain phenomenon by these research institutions too was announced.[30] There were supporters of the meteor hypothesis of the CESS scientists as well. Dr L. Godfrey Louis of the School of Pure and Applied Physics (SPAP)[31] substantiated the meteor theory by explaining the physical and chemical properties of the meteor that could have caused the red rain.[32] Another argument proposed in defence of the meteor theory was that the phenomenon was caused by the Delta Aquarids meteor shower.[33] Thus, the media was performing a dual role at this juncture—as a channel for inter-specialist and public communication of science and as a social agent that broke the black box of science open to bring its internal contradictions and ambivalences into full public view.

Parallel to this, there was an attempt to link the phenomenon of coloured rains with other strange geological phenomena and the earthquakes. Dr Sainudeen Pattazhy, president of the Kerala Environment Researchers' Association (KERA), suggested that the coloured rains were linked to well collapses and landslides.[34] He contended that when landslides and well collapses occurred, methane gas along with ores such as selenium was released into the air and these ores got dissolved in the rain to give it different colours. He also claimed that red rain had been reported from Gujarat after the massive earthquake on 26 January 2001. Dr Pattazhy attempted to challenge the argument of the SALS scientists, pointing out that biological cells had always been present in rainwater, but refuted the possibility for their presence in such large quantities.[35] *Mathrubhumi* was the only newspaper which supported this line of argument, although Dr Pattazhy's proposition first appeared in *Malayala Manorama*. The former newspaper's endeavour was to link the coloured rain to the fault line theory it had been supporting since the well collapses. In its report, *Mathrubhumi* drew a linkage between the coloured rains and the increased tectonic activities in the region.[36] The report pointed out that coloured rains occurred only in a zone stretching between the districts of Kannur and Kollam where the maximum number of well collapses had occurred. The emission of certain gases from the earth into the air due to increased pressure on the bedrock was the cause of coloured rains and the colour change of water in wells and ponds, the newspaper contended. The bubbling of water in wells and ponds was suggested in the report as indicative of the release of gases into the air. The report presented the argument as 'suggested by some geologists', without mentioning their names. It seems that the absence of expert names affected the authenticity of the argument—it appeared more as an analysis the newspaper was making on its own. This indicated a peculiar aspect of the scientific public sphere: providing scientific interpretations was still primarily considered to be the responsibility of scientists. Journalists and other non-scientist actors could either make their own interpretations using the available technical resources, or support one group of experts against another. Emergence of fringe theories within the scientific public sphere happened only on certain occasions, especially when the scientific community across institutions were facing massive erosion of public trust. In the context of the coloured rains, there were still experts from certain institutions with high public credibility analysing the phenomenon, and hence quasi-scientific explanations were yet to be entertained in the scientific public sphere.

In their letter to *Mathrubhumi*, two scientists from the Regional Research Laboratory (RRL)[37] at Thiruvananthapuram linked the phenomenon with similar incidents reported in Kerala half a century earlier, and to the episode of the 'fuming-hill' at Idinjimala.[38] Although this was an expert explanation from two scientists belonging to a prestigious institute, the newspaper found their arguments only worthy to be published in the letter to the editor column. These scientists suggested that the red rain phenomenon was linked to the geological activity in the 'Ring of Fire' zone along the ridge of the 'Indo-Australian-Arabian Plate'.[39] The fumaroles[40] which cracked open the lineaments due to imbalances in the Ring of Fire zone resulted in the emission of gases and fine particles which look like meteor ash. Along with this, seaquakes that caused the explosion of the sediment rocks under the sea had occurred, and released microorganisms and algal cells on the surface of the sea. According to their hypothesis, these factors cumulatively produced red rain and landslides in the region.[41] However, such strange hypotheses did not gain much currency and were not picked up by other actors or were amplified in the media. The media was not very interested in interpretations which were completely strange to the public and not directly linked to the interpretations already being marked as vital to the contestations over knowledge in the scientific public sphere. During the activation of the scientific controversy in each case, a specific 'civic epistemological frame' was developed in the scientific public sphere to decide the relational validity of knowledge claims. The gate-keeping mechanism of the regional press was dependent on these civic epistemological frames.

In the days that followed, a series of conjunctures were extended against the meteor hypothesis from various disciplinary vantage points. The meteor theory proposed by the CESS scientists went against the fundamental laws of physics, a veteran scientist retired from the institute argued.[42] Had there been a meteor explosion, he contended, the dust from the meteor would have spread over a wider area because of the high speed of the meteor, unlike what the CESS scientists suggested.[43] He also rejected the argument that the dust had been washed down by rain three hours after the explosion since it contradicted the scientific understanding of raindrop formation. Having rejected the postulates of the CESS scientists, he proposed an alternative hypothesis.[44] In his opinion, water molecules clustered together around the minute crystals of micro pollutants in the air (aerosols)[45] to form coloured raindrops. According to a study conducted by the Physics Department of the Newman College, Thodupuzha, also, the causative agent was purportedly environmental pollution.[46] They pointed out that the early inferences about the phenomenon were false, for the samples collected by the CESS scientists were not directly from the rain, but from the rainwater that had flowed down the roofs of houses. The physicists' analysis of water samples obtained directly from the rain showed that no elements and compounds usually found in meteors were present in the coloured rains.[47] Instead, they found the presence of algal cells in the samples and they noted that the samples were slightly acidic too. The department also embarked upon further research in collaboration with the Photonics Department of the Cochin University for Science and Technology (CUSAT).[48]

Malayala Manorama published two editorial page essays authored by scientists, which presented new causes for the phenomenon. In the first article, Dr Suresh C. Pillai, a scientist from Kerala working in a European university, denied the phenomenon's tacit connection with the earthquakes, well collapses, and wave formation, and suggested the presence of either chemicals like metal oxides, nitrates, and chlorides or of biological materials from algae, fungi, and lichens.[49] His inferences were based on an analysis of similar instances of and scientific explanations given to coloured rains in the past. Dr K.T. Joseph, former director of the Meteorological Centre, Thiruvananthapuram, also argued on the same lines while suggesting that the phenomenon was the aftermath of industrial pollution.[50] He proposed that the sporadic north-western wind that strikes Kerala had carried along pollutants from the industrial zone in Kochi, consequently causing coloured rain when the pollutants reached the rain clouds. These alternative postulations were either directly critiquing the dominant interpretations for their logical errors and/or referring to established scientific metaphors/narratives such as 'industrial pollution', 'acid rains', or known biological processes.

Such a proliferation of postulations forced the CESS scientists to conduct fresh investigations by factoring in new cases of coloured rain. The weaknesses of the meteor theory in explaining the occurrence of the phenomenon at several other locations as well as the presence of different colours in the rain made such a move inevitable.[51] Acknowledging these shortcomings in their initial assessment, the CESS researchers urged the public through a press release to collect water samples from various locations in the region and send those to the institute.[52] The public response, however, was poor.[53] Therefore, the scientists visited various places where coloured rains had occurred in order to collect water samples to carry out further observations.[54] After a couple of days, the CESS declared that their preliminary inference was wrong, as there were no traces of meteor components identified in the samples.[55] Nevertheless, 14 elements were identified in the chemical analysis and apparently the presence of these elements which are usually seen in biological cells gave the scientists a hint to proceed further with the biological analysis.[56] In the biological tests conducted with the help of the Tropical Botanical Garden and Research Institute (TBGRI) at Palode, the presence of a substantial number of red fungal spores in the rainwater samples initially collected from Changanacherry were identified.[57] The CESS in its press release stated that further research would be carried out in association with the TBGRI in order to determine the origin of the fungal spores, and to understand how they reached the rain clouds, why they did not spread over a larger area, and what species of fungus was involved.[58] From then on the CESS scientists kept on stating that their earlier hypothesis of meteor explosion was only a preliminary and temporary inference made on the basis of mere contextual evidences.[59]

Such a volte-face in the argument catalysed the dispute rather than ending it. Many actors suggested the relationship of coloured rains with other geological incidents in the region. The fear that these phenomena including coloured rains and well collapses foreshadowed an impending catastrophe was raised again at this juncture.[60] The situation was compounded by a new phenomenon of trees unseasonably shedding their leaves in many places.[61] Burn marks on leaves too were widely reported.[62] The CESS declined to conduct research on this new incidence by pointing out that it was the responsibility of agricultural scientists.[63]

The new proposition of the CESS was critiqued by many actors, leading to an explosion in the number of postulations, which were a combination of environmental and meteorological interpretations. Dr Sainudeen Pattazhy alleged that the finding was 'strange', reiterating his earlier proposition of the emission of chromium and selenium ores into the air as causing coloured rains.[64] The argument that the coloured rain was acid rain also resurfaced and the unseasonal falling of leaves was suggested as indicative of acid rain.[65] Some meteorologists illustrated that coloured rains were a harmless phenomenon caused by a diversion of monsoon winds that carried soil particles from the Gulf region to Kerala.[66] The meteorologist who proposed this hypothesis skilfully demonstrated that the high iron content in the dust explained the red colour of the rain.[67] There was also an interpretation that algal cells carried to the monsoon clouds by a whirlwind had in turn caused coloured rains.[68] Another study conducted by a senior scientist from the Meteorological Centre at Thiruvananthapuram proposed that volcanic dust due to the eruption of the Mayon volcano in Philippines was responsible for the red rain in the region.[69] Such an outflow of explanations indicates that the scientific public sphere provided space for scientists and non-scientists to experiment with a pool of explanations already available in public. As mentioned earlier, these propositions were still filtered through the civic epistemological frame of the controversy. Nevertheless, the civic epistemological frame was radically different from and much more flexible than the gate-keeping mechanisms (mainly peer review) within science. The scientific public sphere was open to a larger spectrum of ideas; crossovers across disciplinary boundaries of science as well as the fuzzy borders between science and non-science were hence possible. This function of the scientific public sphere as a 'trading zone' between different cultural and epistemological domains—'epistemological cross-space' (Levina 2009: 109)—became more explicit during the course of the controversy, as we shall see in a while.

As the experts' endeavour to explain the phenomenon of coloured rain thus snowballed into a major public controversy, the ambiguity regarding the science behind the phenomenon spun out of proportion. The situation worsened when the TBGRI who had earlier collaborated with the CESS in the biological analysis of the red rain samples publicly denounced the fungal spore hypothesis. The difference of opinion surfaced a day after the public release of the final report by the CESS. Dr T.K. Abraham, the Head of the Department of Microbiology in the TBGRI who had collaborated with the CESS scientists, rebutted the claim of the latter on the ground that fungal spores were not present in adequate quantities to stain the rainwater.[70] He alleged that the TBGRI was misquoted by the CESS, for the institution had never described the presence of fungal spores as the cause of red rain.[71] Questions about the possible origin of the unidentified biological cells[72] in rainwater and the linkage of the coloured rains with tremors, well collapses, and other geological incidents, thus, remained unanswered even after the CESS scientists' findings.[73] Why the phenomenon occurred exclusively in Kerala also remained unanswered. The ambiguities surrounding the incident in Changanacherry were not resolved either. If the red rain in Changanacherry were caused by fungal spores, how then could the explosion just before the rain be explained? [74]

The debate gradually died out by early September. The coloured rain controversy attained closure without reaching consensus over its possible causes. The negotiations over earthquakes, well collapses, and other geological phenomena had also attained closure by that time without any finalization.[75] A major reason for the abrupt closure was an external factor; the attack on the World Trade Center in New York, the USA, on 11 September 2001 received greater attention in the regional press, and consequently the news value of the unusual geological phenomena declined. The fact that the negotiations regarding the coloured rains themselves was heading nowhere might have also played a role in ending the deliberations. However, the matter did not stop there, as we are going to see in the next section.

'It's Raining Aliens!'[76]

Although the controversy had fizzled out by September 2001, the phenomenon was analytically pursued by two physicists, Dr Godfrey Louis and his doctoral student Santhosh Kumar of the SPAP.[77] They found the phenomenon scientifically interesting and started working on it in 2001. Dr Louis had specialized in experimental solid-state physics, electronic instrumentation, and computer-based instrumentation;[78] Kumar's doctoral research was in solid-state physics.[79] Obviously their current specializations and expertise did not allow them to pursue the phenomenon which had been largely dealt within meteorological and biological sciences. Therefore, they had to develop domain knowledge in these fields to study the cells in the rains.

Louis and Kumar managed to upload their research paper on the alien origin of the cells in an online preprint archive of science in October 2003 called arxiv.org[80] and they were invited to present their findings in a conference on 'Astrobiological Problems for Physicists' organized by the Nordic Institute for Theoretical Physics (NORDITA) in Copenhagen, Denmark.[81] Louis and Kumar linked their investigations to the bandwagon of astrobiology, an emerging interdisciplinary field, and argued that their research on the cells from the coloured rains provided evidence for the theory of Panspermia, proposed by Fred Hoyle and Chandra Wickramasinghe in the 1970s.[82] Based on their studies on light rays from distant stars, Hoyle and Wickramasinghe had argued that life existed in outer space, and comets functioned as vehicles carrying the 'seeds of life'. The modern Panspermia theory[83] thus hypothesized that life began on earth when its 'seeds' reached the planet through comets or meteors.

Parallel to the publication of the research paper as preprint in an online repository, Louis and Kumar had contacted Dr Chandra

Wickramasinghe and his team of researchers at the Cardiff Centre for Astrobiology of Cardiff University, UK, who had been working on the theory of Panspermia and were consequently absorbed into the latter's research network. It was at this juncture that they were invited to present their research in the conference in Denmark. The Panspermia researchers who were not accepted by the scientific orthodoxy found potential in the findings of Louis and Kumar in substantiating their theory and aimed at its greater acceptance among the international scientific community. Followed by this fresh enthusiasm on coloured rains, Louis and Kumar's findings appeared as cover stories in reputed popular science magazines published from the West such as *New Scientist* (UK) and *Popular Science* (USA). In 2006 Dr Wickramasinghe visited Kerala and further collaboration was established between the researchers and the international fraternity of Panspermia theorists. The collaboration was mutually beneficial to both; the theory of Panspermia provided a conceptual framework for the Kerala researchers to work with, and the Panspermia theorists found a potential empirical support to the theory, which was suffering from lack of acceptance as a scientific theory. A scientific paper by Louis and Kumar was published in *Astrophysics and Space Science*,[84] a journal where the proponents of the Panspermia theory used to publish their papers since the late 1970s after a strong decline of their scientific legitimacy.[85] The British Broadcasting Corporation (BBC) also expressed interest in the findings, and its crew visited the researchers in 2006 to make a documentary on their research.[86]

The research of Louis and Kumar began to be reported in the regional media only at this stage, following their acceptance in the Western media. The lower scientific validity accorded to their scientific explanations by the regional scientific community was seemingly a major factor that prevented their early entry into the scientific public sphere. Also their arguments were not in conformation with the civic epistemological frame of the scientific public sphere. Therefore, the researchers tried to get visibility first through an online repository that published non-peer-reviewed research papers to share their research immediately with the peers, before publishing it in peer-reviewed journals. Arxiv. org accommodated their research findings, and accorded them some visibility among the international scientific community. This was followed by the adoption of the strategy that the proponents of the theory of Panspermia conventionally chose—-the route of popular science.

When they could not publish their research in reputed scientific journals, Fred Hoyle and Chandra Wickramasinghe had turned to popular science magazines like *New Scientist*, and newspapers like *The Times* and *The Daily Telegraph* (Gregory 2003: 28). Foyle also wrote popular books like *Life Cloud* (1978), *Evolution from Space* (1981), and *The Intelligent Universe* (1983) about their theory (Gregory 2003). Foyle and Wickramasinghe's research papers also appeared in *Astrophysics and Space Science*, a small, peer-reviewed specialized journal of astronomy, which enjoyed low prestige (Gregory 2003: 35). Louis and Kumar also followed the same strategy: their findings first appeared in popular science magazines from the West, and a documentary film on their research was made by the BBC. This was followed by research publications in arxiv.org and *Astrophysics and Space Science*, the Panspermists' favourite journal. This international media acclaim created ripples in the national media too. There were several popular articles published in the national dailies and magazines,[87] although the regional newspapers continued to be not very open to them, except for *Malayala Manorama* and some television channels.

Malayala Manorama highlighted Louis and Kumar's investigations in 2006, and it coincided with the visit of Dr Wickramasinghe and the BBC team.[88] The Cardiff professor delivered a public lecture at Mahatma Gandhi University followed by a press conference on 30 May 2006. The news reports in *Malayala Manorama* presented the alien cell theory in a spectacular manner. Along with small daily news reports, it published a detailed new story on the research in its Sunday magazine called *Sree*.[89] Quite visibly, a strong collaboration was formed between the newspaper and the scientists, which was indispensable for the latter to attain public visibility, especially in the context of a strong indifference to their propositions from the regional scientific community as well as other newspapers. The regional scientific community's callousness was part of a globally existing scepticism about the scientific status of the Panspermia theory.[90] Their findings grabbed more attention from the international and national media than the regional, although their association with *Malayala Manorama*, the Malayalam daily with the largest circulation, proved to be rewarding.

Dr Godfrey Louis and Santhosh Kumar propounded that the red colour of the rain was not due to fungal or algal spores, nor could it be explained in terms of dust particles from the Middle East. This was argued on the basis of their observations of the rainfall pattern in the region. They pointed out that coloured rain occurred all of a sudden at a specific locality, while rainfall in the other regions was normal. Their study revealed that most of the occurrences of coloured rain concentrated in central Kerala, and a sum total of 124 incidents occurred in the region. The explosion heard by many people just before the first incident of red rain in Changanacherry was a 'sonic boom' caused by a meteoric explosion, they argued. According to the researchers, the red biological cells on the meteor spread all over the atmosphere after the explosion and eventually reached the rain clouds. When the burning meteor was moving fast in the southeast direction, small fragments kept on separating from it and this caused coloured rains at different locations. Their data also showed that 85 per cent of coloured rains occurred within 10 days of its first occurrence in Changanacherry. The red rain continued in the region, Louis and Kumar opined, because of the slow disintegration of the meteor particles in the atmosphere. Such incidents of coloured rain happen all over the world, but the phenomenon was often misconstrued as acid rain by environmental scientists.[91]

Louis and Kumar analysed the water samples using the laboratory facilities of the university as well as of the Sree Chitra Tirunal Institute for Medical Sciences and Technology at Thiruvananthapuram[92] and the Central Marine Fisheries Research Institute (CMFRI), Kochi.[93] The laboratory investigations suggested that the biological cells in the samples were a-nucleate cells devoid of DNA, although cell division was observed. They also found that these cells could withstand high temperatures, germinating even at 300 degree Celsius.[94] The cells had a long life, and survived for more than two and a half years at room temperature. The chemical analysis of the suspended matter revealed the presence of carbon and oxygen in the cells, along with silicon, nitrogen, and hydrogen. Based on these evidences, the scientists argued that the red cells could be a 'strange form of life', which had come from outer space. The news report in *Malayala Manorama* revealed that further research would be carried out in order to prove that these were biological cells with genetic material of an unknown type.[95]

The researchers were actually drawing elements from mutually contending accounts that appeared in the scientific public sphere and incorporating parts of these arguments into their own explanation. Skilfully making use of the resources provided by the scientific public sphere, they succeeded in producing a new interpretation about the red rain. Both the meteor theory and the biological cells theory proposed by the contending groups of scientists during the controversy, thus, neatly fit into the extraterrestrial cell theory.[96] Even the novel component, that is, the 'extraterrestrial nature' of the cells, was de facto not new—Dr Godfrey Louis and Santhosh Kumar carefully mounted their findings as strong evidence for the Panspermia theory, which was premised on the idea of the existence of 'alien cells'. They also successfully situated their research in the interdisciplinary area of astrobiology and highlighted their discovery as 'path breaking' via popular science, while never enjoying acknowledgement from the scientific community. The strong bond that developed between the researchers and *Malayala Manorama* gave them greater visibility in the region, though their discovery failed to animate the scientific public sphere.

The story of these researchers reveals the complex knowledge circuits that exist between scientists in the West and postcolonial locations like Kerala. Louis and Kumar situated the phenomenon as providing strong empirical evidence to the theory of Panspermia. They became internationally recognized by manoeuvring the postcolonial geopolitics of science. At the same time, it was not a centre–periphery interaction in science in its traditional sense. Louis pointed out in a personal interview that if they had handed over the data and the rain water samples to the Western Panspermia theorists at the initial stage, they would have lost the game.[97] Instead of allowing their Western collaborators to take credit, Louis and Kumar themselves worked on the problem with the highly limited scientific facilities they had, being at the periphery of global science. The red rain researchers' sail over the uneven power matrix of global science was made possible by crossing multiple borders: on the one hand, their location at the periphery of global science (as scientists in a postcolony) paradoxically opened up possibilities for them to be part of the international community of researchers who worked on cosmic origin of life. Their geographical location provided them exclusive access to the highly localized unnatural phenomenon of red rain, and this became an invaluable resource for them to negotiate the global power matrix of science.[98] On the other hand, their voyage was to the periphery *within* the centre—the Panspermia theorists' marginal location within science despite being geographically located in the West helped the red rain researchers from Kerala transcend their own marginalities.[99] Notably, this voyage became possible also through a skilful manoeuvring of mass media for the communication of their research.

The findings of Louis and Kumar could have reinvigorated the public controversy, but it did not. The scientific public sphere did not become active, and their findings were paralleled by a closure. Dr Sainudeen Pattazhy and Dr K.C. John, two important actors in the scientific public sphere since the origin of the controversy, opposed the extraterrestrial cells theory, upholding their earlier hypotheses.[100] However, their opposition also received a cold response, mainly because of two reasons. On the one hand, most of the members of the regional scientific community did not oppose Louis and Kumar's research in public, apparently due to an internal consensus over the 'naivety' of the alien cell theory. They seemingly did not want to give extra mileage to the alien cell theorists by arguing against them. The rejection of the life-from-space hypothesis by the international scientific community of astrophysicists and biologists since the late 1970s served as a background for this dismissal.[101] Therefore, no one else from the regional scientific community joined the fleet of Dr K.C. John and Dr Sainudeen Pattazhy. On the other hand, the regional press and the scientific-citizen public apparently lost their interest in maintaining the controversy after three years of the incidents of coloured rains. Thus, the controversy attained closure, and the possible linkage of the coloured rains with other geological phenomena in the region stands invalidated. Some researchers have recently supported the CESS scientists' later argument in a paper they co-authored with scientists from TBGRI—that the red rain contained the spores of *Trentepohlia*, a lichen forming algae (Kumar et al. 2002). Louis and Kumar's argument that the red cells were without DNA too has been challenged (Gangappa and Hogg 2013). Also, new incidents of red rain were reported from Kerala (2011, 2013) and Sri Lanka (2012), leading to some new arguments from both camps of researchers (Nori et al. 2013; Bast et al. 2015).[102] Any conclusive explanation of the strange phenomenon has not appeared till date.[103]

* * *

The story of the coloured rains made explicit the internal ambivalences and ambiguities of modern science. During the controversy, as we have observed in the previous cases also, the internal polysemy of modern science that otherwise remains hidden from public gaze was brought into the open by the regional media. The collaboration of scientists from diverse disciplinary backgrounds, the competition between different professional networks, and the circulation of ideas and things through these channels became publicly exposed during the deliberations. For instance, the collaboration of the CESS scientists with the TBGRI helped the former draw new conclusions by sharing scientific expertise and ideas. The sharing of the rainwater samples and the slides and photographs of the biological cells under microscope helped them coin new hypotheses. Similarly, the proponents of the extraterrestrial cell hypothesis also benefited highly from the global circulation of ideas and things, as we have discussed.

The regional press functioned as a medium for scientists and the public to reflect upon and engage with a wide range of scientific explanations that were proposed, discussed, and falsified or accepted through a complex deliberative process. The scientific public sphere was an important communicative space that enabled conversations between experts from different areas of research and institutional backgrounds. It turned out to be a trading zone for different subcultures of science. Concepts and categories from diverse disciplines were traded between the actors in this contact zone—a process that is not easily permitted within the professional culture of science. The dividing lines between science and non-science as well as the borders between different disciplines, subcultures, and networks were negotiated more flexibly during the public deliberations than within professional science.[104] As we have observed in the case of the alien cell theory, the scientific public sphere constituted by the regional media also functioned as an epistemic repertoire for scientists and non-scientists alike to make new hypotheses and interpretations by tinkering with scientifically less appreciated theories and conjunctures that had appeared earlier in the public deliberation. The strong structural coupling between science and media provided a fresh means for scientists to perform inter-specialist communication outside the conventional modes. This also resulted in the mushrooming of diverse fringe theories that renegotiated the borderline between science and non-science despite these postulations being filtered through the civic epistemological frame of the controversy. The scientific public sphere as a site for expert communication thus offered the non-scientist actors a unique opportunity to engage with science-in-the-making.

Notes

1. 'Changanacherryil "Chuvanna Mazha"', *Deshabhimani*, 26 July 2001, p. 7; '"Chuvappumazha" Apoorva Prathibhasam, Karanam Ajnjatham: Sastrajjnar', *Mathrubhumi*, 27 July 2001, p. 1.

2. *Deshabhimani*, however, published a detailed report of the red rain experience. The report even mentioned who had first noticed the red colour of the rainwater and explained how this personal experience became a shared experience of the people in the locality. See 'Changanacherryil "Chuvanna Mazha"', *Deshabhimani*, 26 July 2001, p. 7.

3. See the reports in the regional newspapers on 28 and 29 July 2001. Red rains had been reported from all over the state but there were comparatively more events reported from the Central Travancore region. Green and yellow rains were also reported, and eventually the regional media started using *varna mazha* (coloured rain) to refer to the phenomenon.

4. They belonged to the Department of Physics and Meteorology in the CESS. Dr Sasi Kumar specialized in atmospheric sciences, and Dr Sampath, in atmospheric and space sciences.

5. '"Chuvappumazha" Apoorva Prathibhasam, Karanam Ajnjatham: Sastrajjnar', *Mathrubhumi*; 'Chuvanna Mazha Indiayil Adya Sambhavam', *Deshabhimani*, 27 July 2001, p. 10; 'Chuvappumazha: Sastrajjnar Parisodhana Nadathi', *Kerala Kaumudi*, 27 July 2001, p. 3.

6. 'Chuvanna Mazha Indiayil', *Deshabhimani*; 'Chuvappumazha: Sastrajjnar Parisodhana', *Kerala Kaumudi*.

7. Personal interview with Dr Sasi Kumar on 27 April 2006.

8. See the news reports of such incidents from Pathanamthitta, Kanjirappally, Meenachil, Palakkad, Kottayam, Ernakulam, and Koothuparampu.

9. See 'Nalu Jillakalil Koodi Chuvappumazha', *Malayala Manorama*, 28 July 2001, p. 1; 'Samsthanathu Palayidathum Chuvappumazha', *Madhyamam*, 28 July 2001, p. 1; 'Chuvappumazha Vyapakamakunnu', *Deshabhimani*, 29 July 2001, p. 6.

10. 'Sample Ethi: Chuvannamazhayude Ulbhavam Anweshikkunnu,' *Mathrubhumi*, 28 July 2001, p. 1.

11. 'Sample Ethi: Chuvannamazhayude', *Mathrubhumi*; 'Chuvappumazha: Bhouma Sasthra Kendram Padanam Thudangi', *Malayala Manorama*, 29 July 2001, p. 1.

12. 'Chuvanna Mazhaykku Karanam Prathyeka Podipadalam', *Deshabhimani*, 28 July 2001, p. 10. However, the report repeated most of the information which appeared on the previous day's newspaper.

13. '"Chuvanna Mazha" Thalkalika Prathibhasam: Malineekarana Padana Vidagdhan', *Malayala Manorama*, 29 July 2001, p. 1.

14. '"Chuvanna Mazha" Thalkalika', *Malayala Manorama*.

15. P.R. Sreemahadevan Pillai, Professor, Engineering College, Palakkad. 'Chuvappumazha Hanikaramallaththa Jaiva Prathibhasam', *Mathrubhumi*, 30 July 2001, p. 5.

16. See, for instance, 'Varnamazha: Sastrathinu Utharam Muttunnu', *Kerala Kaumudi*, 29 July 2001, p. 1.

17. 'Aswabhavika Prathibhasangal Bheethi Padarthunnu', *Deshabhimani*, 29 July 2001, p. 10.

18. 'Aswabhavika Prathibhasangal Bheethi', *Deshabhimani*; 'Sample Ethi: Chuvannamazhayude', *Mathrubhumi*.

19. 'Varnamazha: Paribhramam Vendennu Sastrajjnar', *Kerala Kaumudi*, 30 July 2001, p.1.

20. 'Varnamazha: Paribhramam Vendennu', *Kerala Kaumudi*.

21. 'Changanacherriyile Chuvanna Mazhaykku Karanam Ulkka Pottitheri', *Malayala Manorama*, 31 July 2001, p. 1; 'Chuvanna Mazhaykku Karanam Ulkka Sphodanamanennu Nigamanam', *Deshabhimani*, 31 July 2001, p. 1; 'Chuvappumazha Ulkka Pottitherichathu Moolam', *Kerala Kaumudi*, 31 July 2001, p. 1.

22. 'Chuvappumazha Ulkka Pottitherichathu', *Kerala Kaumudi*.

23. 'Chuvappumazha Ulkka Pottitherichathu', *Kerala Kaumudi*.

24. 'Chuvanna Mazhayekkurichu Bhinnabhiprayangal: Karanam Ulkka Sphodanamennu CESS, Jaiva Kosam Kandethiyennu M.G. Sarvakalasala', *Mathrubhumi*, 31 July 2001, p. 1.

25. Dr A. Chandran assisted Dr K.C. John in the investigation.

26. 'Chuvanna Mazhayekkurichu Bhinnabhiprayangal', *Mathrubhumi*.

27. 'Chuvanna Mazhayekkurichu Bhinnabhiprayangal', *Mathrubhumi*.

28. However, the battle lines drawn at this stage between the biologists and meteorologists did not remain the same forever, as we are going to see later in the chapter.

29. 'Chuvanna Mazha Vellathil Jaiva Kosangalundennu Sthitheekarichu', *Deshabhimani*, 1 August 2001, p. 7.

30. 'Chuvanna Mazha Vellathil', *Deshabhimani*.

31. The School is located in the campus of Mahatma Gandhi University at Kottayam.

32. 'Ulkka Dhoolikalayi, Chuvanna Mazhayayi', *Malayala Manorama*, 1 August 2001, p. 8, editorial page article. Dr. Godfrey Louis turns out to be a crucial speaker later on. See the next section.

33. 'Varnamazha: Uravidam Ulkkakal?, *Deshabhimani*, 4 August 2001, p. 1. This argument was raised by S.R. Prabhakaran Nair, professor of astrophysics at the University of Kerala, Thiruvananthapuram.

34. 'Ulkka Dhoolikalayi, Chuvanna', *Malayala Manorama*.

35. 'Ulkka Dhoolikalayi, Chuvanna', *Malayala Manorama*.

36. 'Chuvappumazhaykku Kinar Idichilumayi Bandham: Bhoojala Vithanathil Ettakkurachil', *Mathrubhumi*, 1 August 2001, p. 5.

37. In 2007 it was renamed as the National Institute for Interdisciplinary Science and Technology (NIIST).

38. S. Suresh Kumar and Dr A.M. Nair, 'Chuvanna Mazha Enthukondu?' *Mathrubhumi*, 3 August 2001, p. 4, letter to the editor. See the discussion on the phenomenon of the fuming hill at Idinjimala in the previous chapter.

39. The term is used here as it has been referred to in the newspaper.

40. A fumarole may be defined as a vent in a geologically sensitive or volcanic zone that releases steam and hot gases.

41. Kumar and Nair, 'Chuvanna Mazha Enthukondu?' *Mathrubhumi*. The same team of scientists explained the phenomena of tremors and well collapses in terms of the 'Ring of Fire' theory in an editorial page article in another newspaper. See Chapter 5.

42. Dr N.G.K. Nair, who was previously the head of the Resource Analysis Wing of the CESS. See 'Chuvanna Mazha: Ulkka Vaadathinethire Ethir Vaadangal', *Malayala Manorama*, 1 August 2001, p. 9.

43. According to him, the minimum speed of a meteor is 17.5 kilometres per hour. See 'Chuvanna Mazha: Ulkka', *Malayala Manorama*.

44. N.G.K. Nair, 'Chuvappumazha Malineekaranam Moolam', *Malayala Manorama*, 2 August 2001, p. 8, editorial page article.

45. According to him these pollutants mainly include laterite crystals that reach the air from soil aided by strong whirlwinds.

46. 'Varnamazhaykku Karanam Anthareeksha Malineekaranamennu', *Madhyamam*, 1 August 2001, p. 9. A detailed report of the study is available at http://drssraman.hpage.com/ (accessed on 27 March 2016).

47. 'Varnamazhaykku Karanam Anthareeksha', *Madhyamam*.

48. 'Varnamazhaykku Karanam Anthareeksha', *Madhyamam*.

49. Suresh C. Pillai, 'Puthiya Prathibhasamalla', *Malayala Manorama*, 2 August 2001, p. 8, editorial page article. He was a faculty member at the Department of Chemistry, University of Dublin, Ireland.

50. K.T. Joseph, 'Pradhana Villain Mazha', *Malayala Manorama*, 2 August 2001, p. 8, editorial page article.

51. The inadequacy of the hypothesis was well accepted by the CESS scientists themselves. See 'Varnamazha: Vividha Sthalangalile Samplekal Sekharikkunnu', *Madhyamam*, 1 August 2001, p. 9.

52. 'Varnamazha: Vividha Sthalangalile', *Madhyamam*; 'Chuvappumazha: Padanathil Janangalkkum Pankedukkam', *Malayala Manorama*, 2 August 2001, p. 1; 'Varna Mazhavellam Adhikritharkku Nalkanam', *Kerala Kaumudi*, 2 August

2001, p. 1; 'Chuvappumazha: CESS Padanam Nadathunnu', *Mathrubhumi*, 2 August 2001, p. 8.

53. Dr V. Sasi Kumar of the CESS research team told me that, contrary to their expectation they received only one useful sample from the public. Personal interview with the scientist on 27 April 2006.

54. See, for example, 'Moonnidathu Chuvappumazha: Kinaridichil Thudarunnu', *Mathrubhumi*, 2 August 2001, p. 11.

55. 'Chuvannamazhaykku Pinnil Fungus: Bhouma Padana Kendram', *Malayala Manorama*, 5 August 2001, p. 1; 'Ulkkavaadam Matti: Jaiva Kosangal Thanne', *Mathrubhumi*, 5 August 2001, p. 1; 'Chuvannamazhayil Fungusum Pathinnalu Moolakangalum', *Deshabhimani*, 5 August 2001, p. 1; 'Chuvanna Mazhaykku Karanam Pooppal, CESS', *Madhyamam*, 5 August 2001, p. 5; 'Chuvappumazhaykku Karanam Fungus Ennu "CESS"', *Kerala Kaumudi*, 5 August 2001, p. 2.

56. 'Chuvappumazhaykku Karanam Fungus', *Kerala Kaumudi*.

57. 'Chuvappumazhaykku Karanam Fungus', *Kerala Kaumudi*.

58. 'Chuvappumazhaykku Karanam Fungus', *Kerala Kaumudi*. See also the technical report that presented the findings of the research team (Sampath et al. 2001)

59. See 'Chuvannamazha: Karanam Vyakthamakathe Sastrajjnar Kuzhangunnu', *Malayala Manorama*, 7 August 2001, p. 9; 'Prakrithi Prathibhasangalil Paribhranthi Vendennu Sastrajjnar', *Madhyamam*, 8 August 2001, p. 5.

60. See, for example, Professor V. Gopinathan, 'Bhoomiyile Albhutha Prathibhasangal', *Deshabhimani*, 7 August 2001, p. 4, editorial page article. The author was the Head of the Department of Geology, Government College, Kasaragode. See also N.P. Radhakrishnan, Pattambi, 'Prakrithi Prathibhasavum Samsthanathinte Thakarchayum', *Deshabhimani*, 17 August 2001, p. 4, letter to the editor.

61. See 'Thrissurum Kasargodum Swayam Bhoovayi Kinar: Kollathu Kootta Ila Pozhichil', *Malayala Manorama*, 5 August 2001, p. 9; see also reports published over the following days. 'Plavu Pettannu Unangi, Ila Kozhichu', *Malayala Manorama*, 6 August 2001, p.1; 'Aryaveppu Unangi Ilakal Kozhichu', *Malayala Manorama*, 11 August 2001, p.1; 'Naalidathu Kinar Thazhnnu', *Mathrubhumi*, 7 August 2001, p.5; 'Ila Pozhium Kalam', *Mathrubhumi*, 8 August 2001, p. 3; 'Aaliyattum Paththekkarilum Ilakal Kariyunnu', *Deshabhimani*, 13 August 2001, p. 3; 'Ila Kozhichilinu Samanamilla', *Madhyamam*, 6 August 2001, p. 7. Reports on leaf fall started appearing in the regional press exactly on the same day that the CESS scientists declared that coloured rains were due to fungal spores.

62. 'Thrissurum Kasargodum Swayam', *Malayala Manorama*; 'Ila Kozhichilinu Samanamilla', *Madhyamam*.

63. 'Ila Kozhichil: CESS Padanam Nadathilla', *Madhyamam*, 7 August 2001, p. 5. Although the CESS did not officially carry out research on the matter, John

Mathai, a scientist from the institute, later on explained the phenomenon as the effect of the electromagnetic waves produced on the surface of earth because of pressure differences on the bedrocks. See 'Kinar Idichilum Varnamazhayum: Asanka Vendennu Vidagdhar', *Malayala Manorama*, 9 August 2001, p. 2.

64. 'Chuvannamazha: Bhouma Padana Kendrathinte Kandethal Vichithram: Paristhithi Gaveshana Association', *Madhyamam*, 6 August 2001, p. 5; 'Chuvanna Mazhaykku Pinnil Jaiva-Raasa Khadakangal', *Madhyamam*, 9 October 2001, p. 10.

65. Findings of the Physics Department of the Newman College, Thodupuzha was quoted by *Madhyamam* in support of this argument. 'Varna Mazhaykku Karanam Amla Vasthukkal', *Madhyamam*, 7 August 2001, p. 5.

66. See 'Varnamazha Asadharana Prathibhasamalla: Kalavastha Vidagdhar', *Malayala Manorama*, 6 August 2001, p. 13. This was opined by Dr C.K. Rajan, Director of the Centre for Monsoon Studies at CUSAT. A similar explanation was reportedly given by the Vikram Sarabhai Space Centre (VSSC) at Thiruvananthapuram. See 'Red Rain: Study Comes Out with New Theory', *The Hindu*, 2 June 2006, p. 8.

67. 'Varnamazha Asadharana Prathibhasamalla: Kalavastha Vidagdhar', *Malayala Manorama*, 6 August 2001, p. 13.

68. K.P. Devadas, 'Chuvannamazha Algaekal Moolam?', *Mathrubhumi*, 15 August 2001, p. 4, editorial page article.

69. The study was conducted by K.K. Sasidharan Pillai. 'Varnamazha: Uravidam Philippines Agniparvatham', *Deshabhimani*, 2 September 2001, p. 1. According to his theory, acids and minerals from the volcano reached the rain clouds in Kerala.

70. 'Chuvannamazha "CESS"ine Vellathilakkunnu', *Kerala Kaumudi*, 6 August 2001, p. 1.

71. 'Chuvannamazha "CESS"ine Vellathilakkunnu', *Kerala Kaumudi*.

72. The argument of the CESS that the biological cells were fungal spores was opposed by hypothetical explanations involving algal cells, lichen spores, and also protozoa. Scientists were not able to identify the species either. See various reports on this regard including 'Chuvannamazhaykku Karanam Protozoakal Anennu', *Malayala Manorama*, 7 August 2001, p. 9; 'Chuvannamazha: Karanam Vyakthamakathe', *Malayala Manorama*; Devadas, 'Chuvannamazha Algaekal Moolam?' *Mathrubhumi*. However, Dr T.K. Abraham continued his research collaboration with the CESS scientists and published a paper that proposed the spores of the alga *Trentepohlia* as the cause for the red colour of the rain. See Kumar et al. (2002).

73. 'Chuvannamazha: Karanam Vyakthamakathe', *Malayala Manorama*.

74. See 'Chuvappumazhaykku Karanam Fungus', *Kerala Kaumudi*.

75. However, this does not imply that coloured rains had stopped by the time. Several incidents of coloured rains were reported in the regional press

after September 2001. For an incident of yellow rain, see 'Manja Nirathil Mazha Vellam', *Malayala Manorama*, 25 June 2006, p. 20. Although such incidents found space in the regional press, no persistent debate followed.

76. This expression is adopted from 'It's Raining Aliens!' *New Scientist* podcast, 4 March 2006, http://www.newscientist.com (accessed on 20 October 2006).

77. Dr Godfrey Louis shifted to the Department of Physics of the CUSAT in October 2006.

78. http://www.education.vsnl.com/godfrey/ (accessed on 20 October 2006). See also https://sites.google.com/site/godfreylouis/ (accessed on 7 December 2015).

79. See his web page, https://sites.google.com/site/asanthoshkumar70/home-1 (accessed on 7 December 2015).

80. See Louis and Kumar (2003), available at http//:arxiv.org/abs/astro-ph/0312639 (accessed on 20 October 2006). The papers appearing in arxiv.org are not peer reviewed.

81. For details of the conference, see http://www.nordita.dk/conference/AstroBio2004/ (accessed on 25 November 2006).

82. Fred Hoyle (1916–2001) was a British astrophysicist. He was the founder of Institute for Theoretical Astronomy in Cambridge, and was also a science policy expert, science popularizer, and science fiction writer (Gregory 2003). Dr Chandra Wickramasinghe was his colleague and former student. Hoyle and Wickramasinghe's early research on astrophysics was well accepted by the scientific community, and their papers appeared in the prestigious interdisciplinary journal *Nature* in the 1960s and 1970s (Gregory 2003). However, by 1978, they experienced a rejection of their papers by *Nature*, and hence started publishing in a peer-reviewed journal of lower prestige called *Astrophysics and Space Science* (Gregory 2003). Panspermia theory was later rejected by the scientific community, and their research began appearing generally in popular media rather than in peer-reviewed scientific journals (Gregory 2003). See Gregory (2003) for a detailed analysis of their life-from-space theory, and its gradual rejection by the international scientific community. For an autobiographical narrative of the turbulent trajectory of his collaborative work with Fred Hoyle, see Wickramasinghe (2005).

83. The theory of Panspermia was first proposed by the pre-Socratic Greek philosopher Anaxagoras.

84. Louis and Kumar (2006), available at http//:arxiv.org/abs/astro-ph/0310120 (accessed on 20 October 2006).

85. Dr Fred Hoyle and Dr Chandra Wickramasinghe were on the journal's editorial board (Gregory 2003: 35).

86. '"Chuvannamazha"yile Jeevante Rahasyam Thedi BBCyum Chandra Wickramasingheyum', *Malayala Manorama*, 27 May 2006, p. 22. The documentary *We Are the Aliens* was aired on 14 November 2006 as an episode of the BBC World's much appreciated science show 'Horizon'.

87. See George Jacob, 'Red Rain: Study Comes Out with New Theory', *The Hindu*, 2 June 2006, p. 8.

88. See '"Chuvannamazha"yile Jeevante Rahasyam', *Malayala Manorama*; K. Tony Jose, 'Niramulla Mazhayil Jeevante Niravu', *Sree* (*Malayala Manorama*), 11 June 2006, pp. 1–5. Also, personal interview with Dr Godfrey Louis and Dr Santhosh Kumar, 23 June 2006.

89. Jose, 'Niramulla Mazhayil Jeevante', *Sree* (*Malayala Manorama*).

90. This view was expressed by many scientists I have interviewed.

91. Personal interview with Dr Godfrey Louis and Dr Santhosh Kumar, 23 June 2006.

92. '"Chuvannamazha"yile Jeevante Rahasyam', *Malayala Manorama*.

93. Personal interview with Dr Godfrey Louis and Dr Santhosh Kumar, 23 June 2006.

94. Personal interview with Dr Godfrey Louis and Dr Santhosh Kumar, 23 June 2006.

95. '"Chuvannamazha"yile Jeevante Rahasyam', *Malayala Manorama*. From the biological point of view, cell division (both mitosis and meiosis) is possible only if genetic material (DNA or RNA) is present in the cell. Contrary to this, the researchers observed cell division in the cells which did not show any sign of the known forms of genetic material.

96. For a similar strategy being employed by Louis Pasteur in developing a very successful theory of microbes, see Latour (1988).

97. Personal interview with Dr Godfrey Louis and Dr Santhosh Kumar, 23 June 2006.

98. Itty Abraham (1998: 44–5) points out this strategy of 'foregrounding … India's "natural" characteristics as a stimulus to scientific ingenuity in the presence of resource constraints' adopted by researchers, while analysing Homi J. Bhabha's work in experimental physics in the early 1950s.

99. See Abraham (2006) for a discussion on the complexities of the postcolonial location in science.

100. See the news reports in *Mathrubhumi* on 7 June 2006.

101. Most of the scientists I interviewed were extremely sceptical of the scientific validity of Louis and Kumar's arguments.

102. K.S. Rajgopal, 'Unraveling the "Blood Rain" Mystery', *The Hindu*, 1 April 2015, available at http://www.thehindu.com/sci-tech/science/unravelling-the-blood-rain-mystery/article7057859.ece (accessed on 8 December 2015).

103. Recently Louis and Kumar have started co-authoring papers with Chandra Wickramasinghe and other Panspermia researchers from the UK. See Gangappa et al. (2010) and Kumar, Wickramasinghe, and Louis (2013).

104. For the boundary negotiations in science, see Galison and Stump (1996) and Gieryn (1995).

7 The Backstage
Science News in the Making

According to the deficit model of public understanding of science, journalists often 'distort' science for a variety of reasons. The media in general and newspapers in particular are considered to be interested in 'sensationalism', and journalists are often blamed for their poor understanding of science. From this point of view of science communication, an 'unruly public' that unnecessarily criticizes scientists and resists scientific and technological advancements is formed due to the poor and partisan reporting of science by the mass media. These hindrances in the smooth communication of science, according to the deficit model, can be surpassed by providing intensive training to the journalists to be skilful science communicators.[1] Similarly, journalists often accuse researchers for the ambivalences in scientific interpretations and expert advice provided to the media and for their cravings for public fame. The public controversies discussed in the previous chapters, nonetheless, reveal an entirely different social dynamics of public communication of science where the regional newspapers are actively involved. Instead of perceiving the scientific and journalistic communities as being always at a confrontational mode in today's mass mediated democracies, this chapter examines their mutual interaction that leads to the production of science news.

The main focus of this chapter is the backstage activities, or, to put it differently, the journalistic production of science news during scientific controversies. By focusing on the production of science news, the chapter explores the structural characteristics of the scientific public sphere and the professional alliances and interactions at the backstage. The first part of the chapter will look at the organization and presentation of science news by examining the discursive strategies employed by journalists and news editors. The following section will discuss the professional interaction between journalists and scientists as well as journalistic collaborations with a wide range of sources of scientific information.

Framing the News

News media employs different discursive strategies to present news effectively to the reader. Discursive strategy can be broadly defined as 'the use of language and linguistic devices to respond to a political situation to alleviate pressure, avoid risk and/or curry favour' (Lee and Lin 2006: 336). During scientific controversies, the newspapers employed different discursive strategies to foreground or erase the internal ambiguity and ambivalences of science. The skilful use of several news genres such as editorials, editorial page articles, letters to the editor, and so on, helped the newspapers stage or present the news according to their editorial policies and politics. An inquiry into these strategies is important for understanding the tacit interests and politics of the newspaper establishments, and it may help us understand how the newspapers acted as active gatekeepers in presenting science news to the reader. The discursive strategies discussed here are not exclusively used in the production of science news, but are a part of the journalistic paraphernalia in general. Here we examine only those techniques and tools the newspapers employed in the production of public deliberation of the scientific controversies.

Newspapers prioritize news according to their preferences and political leanings. The significance of an 'event'[2] is indicated by the space allocated to it in the newspaper. The event either receives more attention in the subsequent days or slowly dies out on account of several factors depending on the alignment of the news. The course of public deliberation over an event is very much linked to its presentation in the media. Various strategies employed by the press to present the news have a crucial impact on the readers in deriving meanings and this was quite evident in the scientific controversies we have examined.

The earthquake controversy clearly depicted a specific pathway in the alignment of news reports. The deliberations began with the portrayal of a particular incident; all the newspapers reported the tremor that struck the region on 12 December 2000. The event was constructed on the basis of the media's use of a set of discursive frames which had been historically available. This repertoire of discursive frames included the fear that Kerala had become an earthquake-prone region, a scientific understanding of the destructiveness of earthquakes, the discourse about the environmental crises the region had been facing, and anxieties about possible disasters. The social memory regarding the tremors that struck the region in the past as well as massive and destructive earthquakes that occurred in other parts of India and the world were readily available as powerful resources for the framing of the event. When the 12 December tremors occurred, all the newspapers presented it as front-page news, supported by a variety of news reports and journalistic essays on the main page as well as other pages. These were supplemented by reports of local experiences of tremors from different districts, explanations and warnings from scientists and disaster experts, reports on the functioning of the government machinery, past events of tremors in the region, and so on. Photographs were widely used to create a visual impact. The sudden staging of the event was evident in the case of the coloured rain controversy as well, following the first incident of red rain.

The public controversy over the clinical trials at the Regional Cancer Centre (RCC), Thiruvananthapuram, radically differed from these controversies in terms of news production dynamics. The news about the clinical trials was first reported as a trivial issue in the local news pages of newspapers, as nobody expected the incident to grow into a major controversy in the near future. The priority accorded to the episode was influenced by at least three crucial factors: First, the fact that the issue was initially aired by a television channel as an exclusive story. Consequently, none of the newspapers wished to promote it or provide unnecessary media mileage to the channel. Second, the RCC was a major research institute and treatment centre in the region with great public trust and appreciation, and its director was a socially, politically, and professionally influential scientist-cum-administrator. He had great moral credence as a pioneer in cancer therapy in the region.[3] None of the newspapers wanted to touch the institution, as it would badly affect the credibility of the newspaper itself—an indication that the reputation of the research institute prevailed over that of the newspaper. Third, the journalists and editors of the newspapers did not perceive it as a significant issue to be presented on the main pages. The dearth of a repertoire of available discourses and social memories linked to medical research may have also hindered the sudden surfacing of the issue in the regional press.[4] The episode was first reported mostly on the local pages of the Thiruvananthapuram edition of the newspapers, and the controversy only gradually became significant enough to be reported in the main pages.

Closure of the controversies and their fading away from the attention of the newspapers also happened in a similar fashion. Closure of public deliberations over the earthquakes and other unnatural geophysical phenomena were partially because of a shift of focus of the regional press to the realm of international politics following the demolition of the World Trade Center towers on 11 September 2001. As a result, reports on the geological phenomena in the region were relegated from the main to the local pages, causing the gradual demise of the controversy. The trajectories of the coloured rain dispute as well as the RCC clinical trials controversy were similar: the news reports decreased in number due to the erosion of their 'newsworthiness' and receded from the main pages to local pages, and then slowly disappeared altogether.[5]

Local pages of newspapers appealed to an important segment of the readership. These pages were set up to create debates on local issues, and also to capture and enhance the local manifestation of an event (Jeffrey 2003). As we have seen in the previous chapters, the local pages were active sites of negotiations. The newspapers often used the local pages in a tactical manner to regulate the debate, and the RCC case stands out as a unique example. Some of the newspapers attempted to prevent the outbreak of the dispute on the clinical trials and sought to restrict the deliberations to the local pages of the Thiruvananthapuram edition in the beginning.[6] The strategy initially helped the newspapers to contain the eruption of the polemic to the locality where the RCC is situated, but they could not escape reporting the development of the controversy in the later phase. Activation of the scientific public sphere in the local editions gradually opened up the debate and the issue snowballed into a regional controversy. The firefighting of the newspapers completely collapsed at this stage and the regional press was forced to report the controversy to a general readership rather than containing the news to the local edition. Although the newspapers had their own vested interests that urged them to hold back the controversy, they completely failed in their attempt because the activation of the scientific public sphere took the controls away from the media.

Newspapers used different techniques to 'amplify' an event when journalists anticipated its 'newsworthiness'. The technique of amplification was widely used to emphasize an issue, thereby increasing its significance and to woo more public attention. When the tremors occurred, the newspapers presented the destruction caused by quakes and the public's experience of the same in order to increase the impact of the incident on the reader. Similarly, when the well collapses and the coloured rains occurred, the local manifestations of the phenomena and the local public's response to it were widely reported. Local incidents and experiences were connected together and subjected to amplification, lifting them to the status of great *regional* importance.[7] The process was not unidirectional—the local public relied on the regional media to make sense of the local experiences by linking them with similar incidents reported from other localities and the experience of people from other locations. They also depended on the media to derive meaning from the scientific explanations and the disputes and ambiguities concerning local incidents and issues. The effective utilization of the regional press as a source of information helped the public analyse local incidents in an erudite manner, bridging their expertise and experiences with the resources provided by the newspapers. This precipitated in local political action, as in the case of the fuming hill at Idinjimala.[8] This strategy helped in building connections among the local experiences to create a sense of solidarity among the newspaper reading citizens, generating a scientific-citizen public of Kerala.[9]

Similarly, during the RCC controversy, the newspapers amplified the issue by drawing in a variety of responses to the clinical trials. The demonstrations and public meetings carried out by different organizations at various places were brought together by the regional press. Similar cases of clinical trials and related events from all over the world were reported and blended with the regional controversy.[10] In a comparable manner, when unusual geological phenomena struck the region, a wide range of similar incidents from various parts of the world were reported in the Malayalam newspapers—this was also a way of familiarizing with unnatural and bizarre occurrences. During the deliberations over the unusual geological phenomena, parallel occurrences of earthquakes and volcanic eruptions from other parts of the world were rigorously reported, whereas in the RCC controversy, public disputes on clinical trials from the West were explicitly integrated with the regional case.[11]

Another kind of amplification technique used in the regional press was the endeavour to bring in a wide range of experts to the scientific public sphere as actors, who were otherwise not directly associated with the dispute. This strategy was effectively employed by *Malayala Manorama* when it appointed its own team of experts to investigate the phenomenon of well collapses in the region. During the earthquake controversy, newspapers made similar attempts by quoting existing studies and the explanations delivered by reputed scientists in other contexts as well as citing research articles.[12] In the same vein, *Mathrubhumi* published the opinion of well-known environmental activists like Medha Patkar and Vandana Shiva in its attempt to suggest environmental destruction as causing increased seismic activity in the region. Such an attempt to bring in actors from different professional backgrounds amplified the event by catalysing the process of deliberation.

The newspapers evoked social memory to construct a legitimate 'past' for the event, as mentioned earlier. This mode of amplification was effectively used during the tremors, well collapses, and coloured rain episodes. Published past experiences of earthquakes and coloured rains were highlighted by quoting old news reports, historical sources, and scientific databases. Many actors also contributed to this process of reinventing the past.[13]

Sometimes the media tried a strategy that was quite the opposite of the technique of amplification. In the case of the RCC controversy, this strategy was widely used in the regional press: let us call it the tactic of 'selective blackout'. Newspapers did not permit certain issues and events they had published to erupt out into a controversy or public debate, particularly when they reckoned that it went against their interests or politics. The RCC clinical trials were an issue that was selectively blacked out in its first phase by the regional press. Competition and rivalry between the newspapers were important factors influencing the news blackout. When *Malayala Manorama* commissioned its own research team to investigate the well collapses, no other newspapers considered it newsworthy, and completely ignored the matter. Similarly, other newspapers blacked out the 'alien cell theory', forcing a closure to the coloured rain controversy. The selective blackout strategy was thus crucial in attaining closure. However, this could not be successfully continued with on many occasions, when the controversy proliferated through other newspapers and media forms, as in the case of the clinical trials. This indicates that the course of deliberation and its closure were not exclusively determined by the media; it was rather a complex process wherein media's role was crucial but not final.

Presentation of the news was very important to attract the readers, and the newspapers used different strategies towards this. Narrative strategies being employed in structuring the news texts hence demand special analytical attention.[14] A key narrative technique used in the news media was the 'montage style' of reporting. Montage is a film editing technique wherein unique footages are spliced together contrastingly to create a unique visual impact. In a similar manner, the newspapers presented two or more different and often contradictory arguments within a single narrative. The technique was often used in the case of scientific controversies to dispute the perspective of experts of greater authority, by counterpoising it with less authentic claims and explanations. For instance, a news report about the recurrence of a geological phenomenon in Kollam district that appeared in *Madhyamam* skilfully employed the technique of montage to delegitimize the CESS scientists' argument that the micro-tremors were harmless. See the text of the news report.[15]

1. The tremors that have been continuing in the district [of Kollam] for more than one week occurred yesterday also at different places.
2. In spite of the expert opinion that the tremors are harmless, there are reports that the district is earthquake-prone.
3. A team of experts will visit the tremor-stricken places again, today.
4. The indications of the earthquake first appeared on the night of the ninth [February 2001]. Water oscillation was observed in the well in the household of ... [name and address of the house owner]. The water level in the well was rising and cracks on the walls of the house were noticed.... By Saturday such signs of earthquake had spread to nearby places....
5. By Monday the situation worsened and the public panicked when indications of tremors appeared in other places too, such as ... [name of the places].

6. The expert team under the leadership of Dr M. Baba, the director of the CESS, visited the affected places.

7. The experts opined that these were not tremors [*Bhoochalanam*] but seismic vibrations [*Bhouma Kampanam*].

8. However, hundreds of houses were damaged in almost all parts of the district in the following days.

9. Wave formation in wells was observed in all these places.

10. A few days ago, those who inhaled the gas emitted from the well of ... [name and address of the house owner] where wave formation had previously been noticed, experienced dizziness.

11. The gas is conjectured to be radon, an inert gas which is not poisonous but is emitted from the earth in certain places prior to an earthquake.

12. However, no tests have been conducted to identify the gas.

13. It was officially informed that scientists [name of the two scientists] from the CESS would examine the damaged houses as well as those wells wherein water oscillation was observed.

14. However, it had been clarified to the representatives of the citizens the other day that there was nothing to worry about the tremors. The public was also informed that the phenomena would continue till the plates rearranged themselves and equilibrium re-stabilised.

15. It has been proved by several studies that the emission of the radon gas is an indicator for earthquake forecasts. In 1966, a similar incident of radon gas emission had been observed in water bodies in China just before an earthquake.

16. Radon gas is commonly found under the earth in those places where radioactive minerals are present in the soil.

17. The emission of radon gas [from the wells] in Kollam district is explained by the presence of *karimanal*[16] in large quantities.

18. Kollam district is already included in the list of earthquake-prone regions published by the National Geographical [*sic*] Research Institute.

The opening sentence is the description of the incident, following the conventional journalistic norms of presenting the main point in the beginning itself (sentence 1). The CESS argued that tremors were not dangerous, but the statement is challenged in the same sentence itself (sentence 2). The scientific source of the counter argument is not tagged anywhere in the text. The next sentence (sentence 3) states that the CESS team would be revisiting the location soon, hinting that the scientists were reconsidering their earlier opinion regarding the harmlessness of the tremors. The next two sentences (sentences 4 and 5) are actually in continuation of the first sentence (sentence 1), further suggesting the gravity of the event, and then the narrative deflects to the details of the first visit of the team of CESS scientists (sentence 6). The scientific explanation of the CESS team is further explored in the next sentence, but without explaining the scientific terms mentioned, enhancing the ambiguity of the scientific argument (sentence 7). The narrative again shifts to the phenomenon (sentence 8) and points out the origin of the new incident of water oscillation in wells (sentence 9). From the next sentence onwards the entire narrative shifts to the emission of radon gas (sentences 10, 11, and 12). The gas emission from the wells is presented as a refutation of the conclusions of the CESS scientists. It is also implicitly insinuated that the CESS scientists were yet to investigate this phenomenon (sentence 12). Further, the source of the scientific explanation regarding the radon gas emission is not given (sentence 11). Then the narrative again cuts to the proposed visit of the CESS scientists (sentence 13). The stubbornness of the CESS scientists is implied in the next sentence, when the report suggests that the experts were reiterating the official stance despite the widespread occurrence of the phenomenon (sentence 14). The news report contrasts the 'rearrangement of plates' explanation proposed by the CESS scientists with the radon gas emission argument in the next sentence by narrating the Chinese experience (sentence 15). A new problematique is then introduced: the presence of karimanal in Kollam district as an indicator of the presence of radon gas (sentences 16 and 17). The concluding sentence of the news report introduces the finding of another scientific institution (NGRI) whose scientific authority superseded that of the CESS, further de-legitimizing the official scientific explanation provided by the CESS (sentence 18).

The narrative structure of the news report shows how the newspaper tacitly destabilized the expert advice put forward by the CESS through the insertion of parallel descriptions. This subversion was accomplished by contrasting different narratives which were often contradictory in nature. The montage technique is applied in the news reports inadvertently: such news reports are often the creation of the editorial desk from a couple of journalistic write-ups. While splicing up, these accounts are rearranged (and often reinterpreted) by the news editors. And the outcome of this editorial job is a montage narrative that acquires a subversive role.

While the news report discussed here is the result of a passive employment of the montage technique, a news report on the earthquakes that appeared as an editorial page article in the *Malayala Manorama* represents a conscious use of the same.[17] The narrative was presented in a mode where three scientists answered the questions posed by the newspaper.[18] The news report was in the form of an interview, each question being followed by three answers. This form of narrative helped create a debate between three lines of arguments, and the readers could compare and contrast the interpretations of the scientists and draw their own conclusions. This method of montage was effectively used by the same newspaper in the RCC controversy too.[19] A related style of montage employed was inter-textual in nature. While the CESS scientists proposed the meteor theory of the red rain, all the newspapers presented counter arguments aligned as parallel texts. Similarly, when many of the scientists argued that earthquakes could not be predicted, the newspapers contrasted it with explanations from those scientists who emphasized the predictability of the earthquakes.

The news narratives were often constructed with reference to a 'public' conceived as the general public of Kerala. News reports normally spoke on behalf of this imagined public who was invoked in the negotiations in a number of ways. In most of the cases the general reader-public of Kerala was invoked as a collective moral authority all the actors were accountable to. In certain instances, victimhood was attributed to the public wherein the public appeared as 'suffering' in various ways. For instance, in the negotiations over the Mullaperiyar dam issue, the general public of Kerala was portrayed as threatened by a bursting dam and the Tamil Nadu government and their scientists were portrayed as the culprits. In many instances, the public was understood as knowledgeable and political, keen to participate in risk governance. In the discussion on the red rain in Changanacherry, the local residents were considered as witnesses of the phenomenon and as providers of first-hand information about the phenomena they had experienced.[20] There were also instances of the local publics being portrayed as lay-experts, posing serious questions regarding science and frequently confronting scientists and government officials: here, the public appears as a political agent to be heard and seriously acknowledged.[21]

As mentioned earlier, news reports were often built around certain 'discourse frames'—that is, to argue a point, the newspapers evoked

already existing, culturally established discourses. Discourse frames
are the 'abstract cultural codes that are understandable to a large and
diverse audience' as defined by O'Mahony and Schäfer (2005: 100). The
discourse frames can be conceived as a set of ideas and terms that had
taken shape over a period of time through a long process of negotiations at various levels. It is through such a long political process that a
discourse becomes solid enough to function as a 'frame' by providing
linguistic tools and categories to formulate a narrative. The 'environmental frame', which had been invoked in the scientific public sphere
quite often, for instance, took shape through a decade of intense political struggles and ideological debates in Kerala around the issue of environmental problems beginning in the second half of the 1970s. While
the risk issues related to the Mullaperiyar dam was deliberated, this
frame served the purpose of conferring legitimacy upon the arguments
of various actors. The frame was widely used both in the context of the
reservoir-induced seismicity (RIS) debate and that of the coloured rains.
The strange geophysical phenomena were understood as consequences
of the increased environmental destruction in the region. The news
reports contending that the coloured rain phenomenon was nothing
but acid rain due to the heavy industrialization of Kochi and nearby
areas shared the environmental frame of discourse. Similarly, the frame
of geopolitical tension between Kerala and Tamil Nadu was rigorously
employed in constructing the news narrative during the Mullaperiyar
dam dispute, as we have seen elsewhere.

The framing of media reports in this manner facilitated the use
of already existing cultural resources to understand the issue under
deliberation in novel ways. When the coloured rain phenomenon was
discussed, the availability of the environmental frame urged the search
for a wide range of explanations to the phenomenon from within environmental science. The act of framing the narratives in this manner
enabled the newspapers to mould the news reports easily as the scientific terms and linguistic style readily became available for use. This
also helped the readers understand the debate with less difficulty. The
casting of the narrative with the help of the linguistic and knowledge
resources available within a discourse frame rendered more effective
communication possible with the readers, and intensified deliberations in the scientific public sphere. Often, the unexpected introduction
of a discourse frame initiated new twists and turns in the course of

deliberation and sometimes accentuated critical interrogation of the dominant scientific interpretations. Discourse frames thus helped the participants in the scientific public sphere manoeuvre the readily available resources to formulate sound and legitimate arguments.

Actors generally employed a 'scientific language' to add legitimacy to their perspectives during public deliberations, as we have observed in the previous chapters. Newspapers extensively referred to scientific sources such as technical papers in peer-reviewed journals, scientific associations, reputed national and international institutions, and individual scientists from outside Kerala. Sometimes an official scientific source was quoted from, as in the dispute over the Mullaperiyar dam, where the CESS was the main source of authentication. There were many instances wherein internationally acclaimed peer-reviewed scientific journals were quoted to substantiate an argument. The reports in the renowned journal *Science* on the RCC clinical trials controversy were widely cited in the newspapers. The appearance of the issue in a leading scientific journal ascribed immense public legitimacy to the allegations raised by the rival scientists from the RCC. In the coloured rain controversy, the alien cell hypothesis of Dr Godfrey Louis and Dr Santhosh Kumar acquired the status of a promising scientific explanation when *Malayala Manorama* informed the public about the researchers' collaboration with Dr Chandra Wickramasinghe and his research group in Cardiff University. Scientists and institutions from outside the state had always been considered as holding greater scientific authority than those from Kerala. Therefore, as we have seen in the case of the controversy concerning the epicentre and intensity of the tremors between the CESS and the research wing of the Kerala State Electricity Board (KSEB), scientists like Dr B.K. Rastogi , who were based outside Kerala, were generously quoted as sources of authentication. Similarly, Chinese and the Japanese scientists were widely quoted to validate the claim that earthquake prediction was not impossible, as opposed to what the regional scientists had suggested.

News Genres

The newspapers skilfully used several news genres in the course of evolution of the controversies. Each genre had its own unique features and argumentative style. The characteristics of the participants in a debate and their arguments varied in nature and status as per the genre. For instance, an editorial is totally different from a letter to the editor in terms of the argumentative style and narrative structure. Editorials of newspapers helped the newspapers to intervene as actors, whereas in the letters to the editor, the readers appeared as actors. The newspapers actively utilized different genres to promote their interests and steer the controversy as per the need of the moment.

All the five newspapers selected for the study effectively used editorials to express their take on the controversies and therefore the genre was largely persuasive in nature.[22] Though often authored by a senior journalist or the chief editor himself, the editorial goes to the press following a discussion at the editorial desk. While some newspapers published only one editorial a day, some others had more than one editorial on different issues on the same day.[23] The editorial is supposed to be in conformity with the general editorial policy of the newspaper. The media owner's interests explicitly determine the orientation of the editorials. All the five newspapers employed this genre effectively to shape the deliberations in the case of the unusual geological phenomena.[24] On the other hand, editorials were rarely brought out on the RCC controversy. The selective blackout performed by the regional press could have been the main reason behind this dearth of editorials on the matter.[25] Many of the newspapers enthusiastically published the readers' responses to their editorials, generally in the 'Letters to the Editor' columns.

Editorials often recapitulated the content of the news reports that appeared on previous days. Since the news reports normally followed the editorial policy of the newspaper that oriented the news in certain direction, it was easy for the newspapers to produce a retrospective account. In this manner, the newspapers developed a specific narrative style for the editorials—a concise briefing of the deliberations so far consumed the first half of the editorial and the other half explicitly articulated the newspaper's standpoint on the issue. When the newspaper published a second editorial on the same controversy, continuity in argument was maintained. Editorials were, thus, an important genre that helped the newspapers orient the news published in accordance with their editorial policy. The genre functioned as the central mechanism that constructed the newspaper as an actor who actively steered the evolution of the controversy.

As the deliberations over a scientific controversy progressed, the newspapers used the editorial page to publish special essays on related topics. The editorial page articles were largely written by a journalist or a scientist when the scientific public sphere was activated. Interviews with key figures involved in the controversies appeared on the editorial page. When the issue under deliberation reached its climax, more than one editorial page article would appear on the same day. Sometimes the newspaper announced a long essay on a specific theme which was serialized on the editorial page, mostly written by a veteran journalist.[26] Such essays usually explicated the nuances of the controversy in support of the viewpoint of the newspaper.

The editorial page articles were rigorously used by the regional press to explain and discuss the scientific aspects of the event. When experts wrote these articles, it often helped them acquire more significance as actors in the scientific public sphere. When the tremors struck Kerala on 12 December 2000, *Malayala Manorama* published five editorial page articles on the different dimensions of earthquakes.[27] Other newspapers also published popular science articles in the same fashion on consecutive days.[28] Sometimes the genre of editorial page essays was used to stoke up the controversy by contrasting the arguments of different actors,[29] thereby suggesting that there was no singular, standard scientific explanation available. The richness of the genre was quite visible in the regional press. Newspapers used the genre to open up debates by revealing different aspects of the scientific controversy under deliberation, and used the editorial page articles, as they did in the case of any other dispute, as an instrument to persuade the readers and to articulate the newspaper's perspective on the matter. Some of the editorial page articles were written in the style of popular science essays and provided basic information about the scientific issues involved, while others were more provocative and argumentative.

Letters from the readers are published by all the newspapers on a regular basis. The 'Letters to the Editor' column of the newspapers was a very important site wherein the individual 'citizen-reader' appeared on his/her own right and intervened in the deliberations. For the newspapers, the column was important as a genre that enabled them to create the impression that they seriously engaged with their readers and acknowledged their right to comment on the news reports. Quite often, the column is used by newspapers to give space to the discarded voices and arguments of those who have been omitted from other news genres (Richardson and Franklin 2003: 184). Frequently, criticism against the newspaper itself was accommodated in the columns to create the impression that the newspaper was open to public criticisms and that it upheld democratic values and journalistic ethics. However, the 'Letters to the Editor' columns are rigorously and carefully edited by the newspapers with reference to their own conception of an ideal reader.[30] As argued by Richardson and Franklin (2003: 184–5),

> [l]etters are selected and edited in accordance with the political identity of the newspaper, the (often only perceived) values and preferences of their buying readership, and other more technical and professional requirements of space and balance. In the selection and placement of letters, newspapers construct debates (or arguments) within and between letters, simultaneously signalling the pertinence of the included letters to the subject being debated and thereby acknowledging and (depending on how the letter is presented) legitimating their contents.

Letters are also frequently created, edited, and rewritten by the editors in compliance with the editorial policy (Richardson and Franklin 2003). Consequently, the editorial choices and preferences play a crucial role in the making of the letters (Pounds 2006).

The genre played a pertinent role in the scientific controversies under study. *Malayala Manorama*, *Mathrubhumi*, and *Kerala Kaumudi* effectively utilized the 'Letters to the Editor' columns as a powerful tool to construct the debate.[31] An analysis of the letters that appeared in these newspapers reveal that the reader was never represented as passive and ignorant. Instead, the letters to the editor were authored by an active reader-public with varied lay-expertise. The large number of letters from medical doctors to express their solidarity with the RCC administration that appeared in the regional press is an example for a professional community's skilful use of a marginal news genre to appear as a pressure group. The nuanced editing of the column enabled the newspaper to create the impression that the citizen-readers' voices were duly spaced in an unmediated fashion. In many occasions, newspapers used the genre to present its own standpoint as the reader-public's general view by carefully selecting and editing the letters. Scientists also frequently wrote letters, articulating their opinion about the dispute. Thus, the genre was used by the regional press as a space to accommodate alternative views on the controversial subjects. In the case of the unusual geological phenomena, the 'Letters to the Editor' column was intensively used to provide fringe theories and to challenge the advice of the official experts. The columns also made professional rivalry within science explicit, when the scientists who had not had an opportunity to express their views on the controversy otherwise, attempted to do so in these columns.[32]

Visual images were frequently used in the regional press to increase the verity of the news and as an explanatory device. Presentation of photographs and illustrations including graphs, diagrams, and maps enhanced the readability of the news. While the RCC case was reported mostly with the aid of photographs of some of the actors or of public demonstrations,[33] in the case of the strange geological incidents a wide variety of visual images were used. These images were collected from different sources and published along with news reports and editorial page articles. While reporting the earthquakes the newspapers published regional maps showing technical details such as the epicentre, lineaments, past earthquakes in the region, and so on.[34] Seismographs of the tremors were presented quite often for the readers' own assessment.[35] When the well collapses appeared, reports on the scientific explanations were accompanied by visual representations too.[36] During the debate on coloured rains, photographs of water samples[37] and microscopic images of the cells in the samples[38] were published. Pictures of the scientists doing experiments and observations also were widely published.[39]

Visual images were a crucial means to communicate the nature of the geological phenomena which had otherwise been strange and unfamiliar to the readers. Photographs of cracks on the Mullaperiyar dam and on the walls of buildings, the collapsed wells,[40] volcano eruptions,[41] devastation caused by earthquakes,[42] and so on, were published. While the debate over the role of the Idukki dam in recurrently triggering seismic activity in the area had been gaining momentum, a photograph of one of the tunnels formed a kilometre away from the dam was published (the image taken from a lower angle showed the tunnel in the foreground, against the gigantic arch dam that filled the frame) in *Mathrubhumi*, to suggest the precariousness of the situation and to woo public attention.[43] The presence of the local public in some of the visual images was significant; the members of the affected household as well as neighbours appeared in the picture frame.[44] Some of the photographs depicted scientists investigating the cracks with the local residents observing them 'doing science'.[45] Interestingly, photographs quite often depicted women among the local publics, in contrast to their total absence from other genres of news as representatives of the scientific-citizen public.[46]

Nevertheless, certain sub-genres were never effectively utilized by the regional press during the risk controversies. Cartoons were a good example. Although cartoon had been a crucial genre in the regional press, there were none published on the scientific controversies we have examined.[47] A few cartoons, however, evoked the tremors and the well collapses metaphorically to articulate political themes.[48]

Professional Interaction and Mobilization of Resources

The scientific public sphere was radically sustained by the alliances and resource mobilization at the backstage, which is the focus of this section. Newspapers relied on various sources to collect resources for the production of science news. These resources included scientific explanations, statements of experts, press notes, photographs, graphs, maps, readings from scientific instruments, specimens, raw scientific data, scientific papers, media reports from the West, and so on. Individual scientists as well as spokespersons of scientific institutions were important sources of scientific information for the journalists. Journalists' practice of seeking technical advice on controversial topics simultaneously from experts who belonged to different institutions and research fields often became crucial to the development of the controversy. Reporting of multiple and contradictory interpretations were important in catalysing public deliberations.

Each journalist/newspaper had his/her regular and sometimes exclusive sources of scientific information too. Whenever a geological phenomenon occurred, reporters contacted experts and scientific institutions for scientific data as well as interpretation. For instance, when the tremor occurred on 12 December 2000, all the newspapers approached the CESS, an institution that had great credibility as the most authentic source for information on earthquakes in Kerala. The institute was expected to provide the 'official' explanation since it was involved in studying earthquakes and the tectonic structure of the region. Experts were also personally approached by the news reporters for detailed information and scientific explanations. Sometimes this led to the scientists blaming the journalists and the press for 'distorting' what they had originally stated, although there was generally a good degree of understanding between the two professional groups. This was pointed out by Dr C.R. Soman, who turned out to be an important participant in the RCC controversy. He became involved in the dispute only after journalists contacted him because of his expertise in health-related issues. He also felt that the media was approaching him due to his critical views of the medical establishment. He pointed out that the journalists interviewed him mainly to get a 'dissent note'. [49] It is notable that Dr Soman was quite aware of the role of 'the scientific counter-expert' he was supposed to play (Neidhardt 1993: 344). There was perfect professional understanding between him and the news reporters although in some contexts mutual misunderstandings and political differences created breakages in professional collaboration. During personal interviews two prominent scientists from the CESS criticized *Mathrubhumi*'s journalists for their 'distortion' of scientific information. They hinted that they began avoiding reporters from the newspaper after a few bad experiences of 'distortion' of their explana-tions. They also accused that after this incident the newspaper started taking a stand against them and the CESS in the scientific controversy.[50] The RCC controversy made explicit the rivalry between the two alli-ances in which journalists and scientists were part of. There was little collaboration between researchers and journalists belonging to the opposite camps.

The process of gathering information from the experts culminated in the discursive construction of the event. Information collected from the experts was translated into a suitable narrative style by the jour-nalists and editors. What the expert 'originally' meant to say thereby acquired a totally different set of meanings once mediated. An actor in the scientific public sphere was not simply an 'individual out there', but a discursive construction. He/she occupied a nodal status in a web of relations and translated the interests of others into his/her interests and expertise in the process of being an actor on the front stage.[51] Dr C.R. Soman, a powerful participant in the RCC clinical trials controversy, revealed this social dynamics of being an actor by disclosing that he had been contacted by several people who found themselves 'voiceless': some of them were professionally attached to government institutions and therefore did not want to appear in the newspapers.[52] Some others found themselves incapable of establishing links with the journalists, and hence shared the resources they had with him. Dr Soman acknowledged that he was representing these invisible people, although he processed the resources in his own way in order to develop his opinion on the issue as a counter-expert. At the same time, some people urged him to withdraw from the controversy.[53] It indicates that he had to shape his stance critically engaging with a wide range of interest groups who found him suitable to voice their concerns. Experts also represented the interests of scientific institutions, civil society organizations, and different professional groups. Their views were crucially informed by their social location, professional status, and location within the scientific field (Bourdieu 1975). The Mullaperiyar dam issue is a good example that revealed this aspect. 'Being a scientist working for the Kerala government', Dr C.P. Rajendran of the CESS found himself creatively taking into account 'regional concerns' as well as the 'objective realities' while engaging in the issue and countering the arguments of the scientists from Tamil Nadu.[54]

Press conferences of the expert teams appointed by the government as well as those conducted by the scientific institutions and individual scientists were a valuable source of information for the journalists. Press notes released by the scientific institutions were also considered authentic sources of information. The press conferences functioned as a deliberative space in its own right and made the interaction between scientists, government officials, politicians, and journalists possible.[55] Researchers convened press conferences with an intention to regain public trust and end the dispute, but as in the RCC controversy, this strategy often backfired, forcing them to divulge further information in public.

Reporters also approached scientists who were not directly involved in the controversy, as we have seen earlier. Sometimes, individual scientists who disagreed with the interpretations of major scientific institutions and reputed scientists approached the reporters to provide their own counterarguments. In the case of the RCC controversy, those scientists who differed from their colleagues contacted mediapersons to provide information on the matter, and it was this move of the rival scientists that conferred on this event the status of a scientific controversy. As we have seen earlier, newspapers frequently reported such alternative accounts from less authentic sources.[56] Whenever an official team of experts was constituted to probe the scientific aspects of the issues under debate, journalists travelled with the team to obtain first-hand information on the matter. This proximity helped them do in camera reportage of the encounter between scientists and other actors. As we have seen in the previous chapters, there were incidents reported in the newspapers wherein the local public confronted the scientists visiting the affected localities for investigations. Sudden eruptions of such deliberative moments were well captured by the journalists when they travelled with the expert team.

Scientific papers and reports prepared by scientists from India and abroad were referred to by many journalists. Newspapers often relied on scientific journals and popular science magazines in the making of science news. When the RCC clinical trials controversy was deliberated, the regional press brought in related incidents reported in the Western media. Debates in the Western media were passionately followed in order to enrich the deliberations in the scientific public sphere. In the context of the coloured rain controversy, the theory of Panspermia published in Western popular science magazines informed the regional debates. Research of Chinese and Japanese scientists on earthquake prediction was regularly followed by the journalists. In the RIS debate, studies in the context of other dams like Koyna (Maharashtra) were often invoked. The internet was an important source of information for the journalists.[57] There were many instances of scientists too relying on the internet to collect preliminary information on a scientific controversy. For example, the CESS scientists who proposed the meteor theory collected information from the internet regarding similar incidents of coloured rains reported in the past in different parts of the world.[58] Some researchers who lacked professional acknowledgement from their peer group used the internet to introduce their research to a wider audience during the controversies.[59] The presence of the internet as an easily available source of scientific information for the scientific-citizen public contributed to the erosion of scientists' authority over interpretation to some extent.[60]

None of these Malayalam newspapers had a science and technology (S&T) beat like some of the national dailies.[61] However, there were journalists in the regional press who had kept a tab on certain areas of scientific research although they were not exclusively science correspondents.[62] When an issue related to their area of interest emerged, these journalists played an active role in writing news reports as well as participating in the formulation of the editorial policy of the newspaper.[63] Most of these journalists had their basic degrees in various sub-disciplines of science.[64] Such an opening of the gates of journalism to science graduates in Kerala was apparently a trend that started in the 1990s, parallel to the emergence of risk controversies.[65] The majority of the regional newspapers had a reporter at Thiruvananthapuram, the state capital where most of the prestigious scientific institutions are located, who was assigned to contact these institutions and the scientists when a scientific dispute emerged.[66]

Another important factor regarding the journalistic production of science news was the existence of associations and networks which helped the regional press in effective science reporting. Although the newspapers generally accessed multiple sources of scientific information, special networks and channels too were availed by each newspaper to gather scientific resources. Some journalists and newspapers had privileged access to certain scientific institutions and scientists. The regional press generally relied on the CESS scientists for 'authentic' information on tremors and well collapses, but *Mathrubhumi* depended on the Kerala State Groundwater Department (KSGD) research team led by Dr K.V. Mohanan as well, as an alternate source. The RCC clinical trials controversy had at its heart the rival coalitions formed between scientists and journalists.

Privileged access to certain sources of information and unique alliances between certain scientists and journalists had a wider impact. Both the scientists and the journalists benefited from such symbiotic associations. While such associations helped the journalists and newspapers to introduce new twists and turns into the deliberations by accessing exclusive information, it was rewarding to the scientists too, as they could push their arguments more effectively into the scientific public sphere with the aid of the journalists. The association with journalists helped scientists acquire public attention and thereby more social recognition and government funds for their research. More public recognition through media also meant their recurrent inclusion in expert committees nominated by the government, which further increased their public and peer recognition, although sometimes the media exposure backfired and resulted in lesser credibility and greater negativity.[67]

We have already seen how new practices like the media appointing their own expert teams took shape during the public controversies.[68] There were also other interesting aspects linked to the professional symbiosis that influenced the knowledge production practices in science: at certain moments, journalists and scientists worked together in scientific data collection and analysis. *Mathrubhumi* collected data regarding collapsing wells from all over Kerala with the help of its reporters, under the coordination of journalist Sasidharan Mangathil and the CESS scientist John Mathai.[69] The vast journalistic machinery of the newspaper was, thus, used by scientists to collect scientific data. The same newspaper played a crucial role in collecting preliminary information about coloured rains from different parts of the region too with the help of their reporters, when the direct call of the CESS scientists to the public to provide samples failed.[70] I was told that the information collected by the newspapers was of immense help to the scientists in both instances.[71] In a scientific research paper on the unusual geological incidents in the region, the authors duly acknowledged their dependence on daily newspapers for collecting information. The researchers pointed out that '[i]nformation on some of the incidents has been obtained from district authorities and *through Daily Newspapers after proper evaluation*' (Singh et al. 2005: 75).[72] However, note the authors' immediate addition that the data collected from the media was subjected to scientific scrutiny before being used, indicating the authors' fear of peer group disapproval of their novel method of data collection. Personal interviews with scientists revealed that on many occasions, they gathered primary information from newspaper reports.[73] There were also occasions where the media reports provided the basic details about different geological phenomena to the experts. The scientific research system in Kerala, which was poorly staffed, necessitated this compensatory mechanism of drawing upon the media sources. These unique instances of medialization of science showed that the science–media relationship was much deeper than the common rhetoric of 'the media distorting science' and 'the bad scientists going to the media'.

At the same time, such complaints about the other professional group recurrently appeared in the arguments of the scientists and journalists whom I interviewed. It was interesting to see that scientists and journalists associated with each other in diverse creative ways, but used the rhetorical language of the deficit model when they were asked to comment on the other professional group. Scientists often blamed the incapability of journalists in understanding their scientific arguments accurately. The journalists were also accused of sensationalism. Some scientists warned against the 'pseudo-scientists' who approach the media for publicity. Journalists were blamed for promoting these pseudo-scientists. On the other hand, completely oblivious to the internal polysemy of science, journalists alleged that the scientists of Kerala failed to produce 'a single, coherent, and legitimate explanation' of the unusual natural phenomena witnessed in the region, indicating that they employed a positivist understanding of science while blaming scientists. Scientists were accused by the journalists of issuing contradictory statements and were reproached for their craving for popular reputation by appearing in the media. The social dynamics of journalistic production of science news and the professional interaction between scientists and journalists at the backstage strongly indicates that these allegations were largely rhetorical and did not reflect the complexities of the backstage dynamics.

✳ ✳ ✳

As we have discussed in this chapter, the scientific public sphere operated on a sophisticated discursive mechanism that made possible deliberations on its front stage. The journalistic production of news had a significant role in keeping the public controversy alive and kicking. The controversies analysed in the book showed that the public deliberations during the controversy were not entirely regulated or taken over by any single agent, be it the experts, scientific institutions, the journalists, the newspaper establishments, or the government. Since there were several newspapers involved in the constitution of the scientific public sphere, each of them presented science news in unique ways, influenced by the newspaper's political perspective, readership dynamics, and other particular characteristics. The professional competition between the newspapers for increasing their readership and thereby attaining more revenue from advertisements was, for example, a significant factor behind their eagerness in publishing exclusive news reports that became crucial in the development of public controversies over science. The everyday presentation of news reports in each newspaper had a deep impact on the deliberations in the other dailies the next day, affecting the very course of the public controversy. The journalists and news editors employed different techniques and tactics to create special effects during risk controversies, and certain news genres were creatively used. The deliberations in the scientific public sphere were incessantly being remoulded through the dialogical coexistence of the regional newspapers.

The discursive mechanisms of the scientific public sphere examined in this chapter reveal that a complex network of diverse actors was rampant at the backstage and it was through this heterogeneous network that the resources for news production were mobilized and the front stage managed. Professional association between the journalists and scientists was decisive in shaping the production and presentation of science news. Scientists and journalists as two unique professional groups had their own concerns, roles, and expertise, and their symbiotic relationship actively shaped the production of science news. Sometimes their professional relationship during the risk controversies created new practices, such as journalists providing scientific data to researchers for analysis and newspapers commissioning their own research teams to provide expert advice. Sometimes, the professional alliances broke down or became weak due to a variety of reasons, creating crises for backstage management. Such crises, however, contributed to the shaping of the scientific public sphere. The backstage dynamics of the scientific public sphere thus revealed many unique aspects of medialization of science during the risk controversies.

Notes

1. At the same time, media training for scientists is not a common practice.

2. Here the terms 'incident' and 'event' have distinct connotations. The former denotes a 'happening' that occurs at a site outside the scientific public sphere, a natural phenomenon for example. The latter is defined as the discursive construction of a happening within the scientific public sphere. The emergence of an event in the scientific public sphere and the deliberations triggered off by it presupposes a complex discursive process.

3. For more on his contributions to cancer research, see Nair (2013).

4. It is quite interesting that past experiences were hardly invoked by the newspapers in connection with the clinical trials controversy unlike in other controversies. This indicates that contrary to earthquakes, the incident of the clinical trials was a fresh experience for the public of Kerala.

5. The journalists who were interviewed generally suggested that the demise of an issue occurred when the readers lost their enthusiasm. When the deliberations reached this stage of reader disenchantment, the newsworthiness of the event also declined.

6. See the discussion on 'selective blackout' discussed later in this chapter.

7. In this sense the activity of reading newspapers was an active, everyday process of producing Kerala as a 'region'.

8. See Chapter 5 of the book.

9. For a discussion on similar instances of the media bridging isolated hamlets and stranded individuals and providing 'emotional support and companionship' to the public and thereby creating a community in the context of disasters, see Perez-Lugo (2004: 213).

10. See Chapter 4 for a detailed discussion on the clinical trials controversy that appeared in *The Washington Post*, and the way it was linked to the RCC dispute.

11. The everyday production of the region was, therefore, achieved through the accessing of the global as well as the local. Their readers were hence imagined as 'cosmopolitans' by the newspapers.

12. For example, newspapers quoted the study of Dr Harsh Gupta many times in different contexts in support of the RIS theory. Research of Chinese scientists was quoted quite frequently to argue that earthquake prediction was not impossible.

13. See, for example, the letter to the editor published in *Mathrubhumi* in which a reader recollected an incident of red rain that occurred fifty years ago. V.A.N. Unnikrishnan, Perumpavoor, 'Chuvappumazha Anpathu Kollam Munpu', *Mathrubhumi*, 31 July 2001, p. 4. *Malayala Manorama* republished a report on an earthquake that originally occurred on 29 July 1899 in its 'Hundred Years Ago' column. See 'Bhookampam Bhayannu Naaduvittu', *Malayala Manorama*, 19 February 2001, p. 8.

14. The textual analysis here is inspired by methods of discourse analysis in general. For research on public engagement with science and technology (PEST) employing a discourse analysis framework, see Calsamiglia and Dijk (2004); Carvalho (2007); Cook, Robbins, and Pieri (2006); O'Mahony and Schäfer (2005); Yamaguchi and Harris (2004). For a detailed introduction to discourse analysis of media texts, see Dijk (1988).

15. 'Kollath Bhoochalanam Thudarunnu: Vidagdha Sangham Innu Veendumethum', *Madhyamam*, 17 February 2001, p. 1 (author's translation).

16. Karimanal is the beach sand that contains minerals such as ilmenite, rutile, zircon, monazite, leucoxene, sillimanite, and garnite. Many of these minerals are radioactive. This naturally radioactive mineral sand is mostly found on the seashores in the Alappuzha and Kollam districts of the state. See Abraham

(2012) for a detailed study on the biopolitical and geopolitical consequences of mineral sand mining in Kerala.

17. 'Bhoochalanam: Asanka Venda: Enkilum Karuthal Venam', *Malayala Manorama*, 19 January 2001, p. 8, editorial page article.

18. The scientists quoted were Dr Kusala Rajendran of the CESS, Dr B.K. Rastogi from the National Geophysical Research Institute (NGRI), Hyderabad, and Dr Saumitra Mukherjee from Jawaharlal Nehru University (JNU), New Delhi.

19. 'Nirodhicha Marunnukal Pareekshichuvo?' *Malayala Manorama*, 1 August 2001, p. 4, editorial page article. In this report the arguments of Dr M. Krishnan Nair and Dr C.R. Soman were contrasted. This method of montage appeared in the genre of editorial page articles.

20. The newspapers frequently and painstakingly reported how the public experienced an incident of earthquake, well collapse, or coloured rain.

21. The fuming hill incident at Idinjimala (Chapter 5) is the best example.

22. The persuasive nature of editorials is noted in E. Le, 'Active Participation within Written Argumentation: Metadiscourse and Editorialist's Authority', *Journal of Pragmatics* 36, no. 4 (2003): 687–742, cited in Lee and Lin (2006: 335).

23. *Malayala Manorama*, *Kerala Kaumudi*, and *Deshabhimani* generally published only a single but lengthy editorial every day. *Mathrubhumi* and *Madhyamam* used to publish two short editorials per day.

24. Approximately eleven editorials on the issue were published by *Mathrubhumi*, while *Malayala Manorama* published eight. *Deshabhimani* published six editorials and *Madhyamam* and *Kerala Kaumudi* published four each.

25. Only *Mathrubhumi* and *Kerala Kaumudi* published one editorial each in connection with the controversy.

26. The series of articles published in *Mathrubhumi* authored by Sasidharan Mangathil is an example. See Sasidharan Mangathil, 'Keralam Kulungumpol', *Mathrubhumi*, 17 February 2001 to 23 February 2001, p. 4.

27. The main essay was written by a journalist, describing the tremors by situating them within the geological understanding of the region, utilizing data from different scientific sources. The essay discussed the collapse of the belief about a safe 'South Indian Shield', the RIS thesis, and the activation of fault lines in Kerala. See B. Ajith Babu, 'Chalana Kendram Periyar-Idamalayar Bhramsa Meghala', *Malayala Manorama*, 13 December 2000, p. 8. The second essay was written by two reputed earthquake scientists from the CESS, giving the scientific explanation for earthquakes, with an intention to ward off public misunderstandings about this natural phenomenon. Dr C.P. Rajendran and Dr Kusala Rajendran, 'Bhoochalanam Parakkettukal Uranju Neengumpol', *Malayala Manorama*, 13 December 2000, p. 8. These two main essays were flanked by three small articles. The first one explained the purpose and use of the Richter scale,

while the second one provided security tips to escape from earthquakes, and the last one was a box-news item within the journalistic article that described the ability of animals to understand oncoming earthquakes. See 'Richter Scale: Bhoochalana Vyapthiyude Alavukol', 'Bhoochalanam Vannal', and 'Mrigangal Neraththe Ariyunnu', all in *Malayala Manorama*, 13 December 2000, p. 8.

28. A rough estimate shows that more than fifty-seven essays have been published in the regional press in connection with the first case unlike the RCC controversy, in which case only a few essays appeared on the editorial page.

29. See the discussion on montage techniques earlier in the chapter.

30. Sometimes, the 'Letters to the Editor' column is edited in such a way that different editions of the same newspaper on the same day had different versions of the column. This edition-specific editing of the column helped the newspaper to include more letters and to be more sensitive to the local readership. Personal interview with D. Sreejith (Staff Correspondent, *Mathrubhumi*), 11 December 2006.

31. In connection with the controversy over the unusual geological phenomena, more than seventy letters had been published in the regional press, of which almost 90 per cent appeared in these three newspapers. *Malayala Manorama* published twenty-five letters, while *Mathrubhumi* and *Kerala Kaumudi* published eighteen letters each. In the RCC controversy, *Kerala Kaumudi* published twelve letters and *Mathrubhumi* six. The other three newspapers under analysis either published only a nominal number of letters or completely ignored the genre.

32. For example, see the letter from a former scientist of the CESS, criticizing the CESS scientists. Dr M. Manikantan Nair, 'Bhoochalanathil Kulungunna Sastrajjnar', *Mathrubhumi*, Thiruvananthapuram,13 January 2001, p. 4, letter to the editor.

33. See for example the photographs of Dr Krishnan Nair and Dr C.R. Soman in 'Nirodhicha Marunnukal Parisodhichuvo?' *Malayala Manorama*. Mr Gopal's photograph was published along with the report 'Guinea Panni Aayath Aarkku Vendi Ennariyathe Gopal', *Madhyamam*, 3 August 2001, p. 5. The picture of the public demonstration of the RCC Protection Forum appeared as part of the news report 'RCC Samrakshikkan Manushyachchangala', *Malayala Manorama*, 25 September 2001, p. 2. The photograph of one of the public interaction sessions of the Parekh Commission also appeared in the press. See *Mathrubhumi*, 10 August 2001, p. 13.

34. For the regional map of Kerala that depicted such details of the earthquake, see 'Innalaththe Bhoochalanam', *Malayala Manorama*, 13 December 2000, p. 1; 'Keralathil Bhoochalanam, Paribhranthi', *Mathrubhumi*, 13 December 2000, p. 1. For maps showing lineaments, and so on, see 'Bhookampa Sadhyathakalum Munkaruthalukalum', *Mathrubhumi*, 31 January 2000, p.4; Sasidharan Mangathil, 'Bhoomiye Kulukkunna Bhramsana Mekhalakal', 20 February 2001,

p. 1. See the map of the seismic zones in India in 'Bhookampam: Adiyanthira Surakshanadapadi Anivaryam', *Madhyamam*, 28 January 2001, p. 4.

35. See the seismographs of the tremors obtained from the seismic station of the CESS at Peechi. *Malayala Manorama*, 13 December 2007, p. 1; 'Veendum Sakthamaya Bhoochalanam, Palayidathum Nasanashtam', *Malayala Manorama*, 8 January 2001, p. 1.

36. See, for instance, Sasidharan Mangathil, 'Keralathil Bhoovillalukal Sajeevamakunnu: Sastrajjnarkku Asanka', *Mathrubhumi*, 27 July 2001, p. 1, for the visual representation of well collapse occurring due to pressure.

37. See the image of the red rain sample collected from Valanchuzhi in the Pathanamthitta district in 'Palakkattum Kottayathum Chuvanna Mazha', *Mathrubhumi*, 28 July 2001, p. 1.

38. See 'Chuvanna Mazhayekkurich Bhinnabhiprayangal', *Mathrubhumi*, 31 July 2001, p. 1.

39. See, for instance, the photographs of the team of scientists appointed by *Malayala Manorama* visiting the collapsed wells and discussing the subject with scientists and other actors that appeared in the newspaper on 12, 13, 15, and 17 August 2001. Also see the team of scientists from the KSGD examining the collapsed wells at Pavangadu, 'Kinar Aprathyakshamakal Bhoomiyude Puramthodile Chalanam Moolam Ennu Padanam', *Mathrubhumi*, 7 July 2001, p. 1. In the context of the Mullaperiyar dam issue, photographs of the research teams from Kerala examining the dam repeatedly appeared in the newspapers. For example, see 'Mullaperiyar Daminu Villal: Bhookampam Moolamennu Samsayam', *Mathrubhumi*, 16 December 2000, p. 1, lead article. See 'Chuvappumazha: Sastrajjnar Parisodhana Nadathi', *Kerala Kaumudi*, 27 July 2001, p. 3, for the CESS scientists examining the samples of red rainwater at Changanacherry.

40. See, for example, 'Neeramankarayilum Kaladiyilum Randu Kinarukal Idinju Thaanu', *Malayala Manorama*, 10 July 2001, p. 2; 'Kinarukal "Aprathyaksha"mayath Anweshikkan Mukhyamanthri Nirdesichu', *Mathrubhumi*, 18 July 2001, p. 1; 'Thiruvallayilum Kinar Thannu', *Deshabhimani*, 18 June 2001, p. 8.

41. See, for instance, the photograph of Mount Etna in Europe, which showed the pre-eruption fuming of the volcano. *Madhyamam*, 23 July 2001, p. 8. This photograph appeared in the newspaper in the context of the incident of the 'fuming' of the hill at Idinjimala. See also the photograph of the Mayon volcano in Philippines, which was argued as causing the coloured rains in Kerala. 'Varnamazha: Uravidam Philippines Agni Parvatham', *Deshabhimani*, 2 September 2001, p. 1.

42. For a photo feature on the Gujarat earthquake, see *Malayala Manorama*, 29 January 2001, p. 12. All the newspapers had widely covered the event through visual images.

43. *Mathrubhumi*, 13 July 2001, p. 1.

44. See, for example, *Malayala Manorama*, 28 August 2001, p. 11; 'Aaru Kinarukal Koodi Aprathyakshamayi: Kannuril Kulam Thaanu Poyi', *Mathrubhumi*, 17 June 2001, p. 1; *Mathrubhumi*, 14 March 2001, p. 2; 'Kaviyooril Veedu Mannil Thazhnnu', *Deshabhimani*, 8 July 2001, p. 7; 'Veedinu Chuttum Vellam Pongiyath Albhuthamayi', *Madhyamam*, 14 March 2001, p. 3; 'Pettannathaa, Rodinu Naduviloru Kuzhi', *Kerala Kaumudi*, 22 June 2001, p. 10. However, when disasters like the Gujarat earthquake were portrayed, the focus of most of the photographs was the victims. The intention was to create a deeper impact of the intensity of the disaster on the readers.

45. See, for instance, the picture that showed the scientists from the CESS collecting water samples from the coloured rain in Changanacherry. Local people were seen in the photograph as curiously watching what the scientists were doing, invoking the colonial anthropological photographs of the natives actively observing the anthropologist. See *Kerala Kaumudi*, 27 July 2001, p. 3.

46. See, for instance, *Malayala Manorama*, 13 August 2001, p. 2. In this photograph showing the expert team appointed by the newspaper examining a collapsed well, a woman (apparently she was the 'housewife') was present as a curious observer. Similarly, in another photograph of a well collapse, a woman was shown as pointing her finger towards the collapsed well, while another woman was looking into the well. In the news report that was given with the image, a young woman was mentioned as being entrapped in a collapsed well. 'Kottathalam Idinju Yuvathi Kazhuthattam Bhoomiyil Thazhnnu', *Mathrubhumi*, 15 August 2001, p. 1.

47. Cartoons often depict political themes in the Malayalam press. Sharmila Sreekumar's analysis of cartoons on the risk of HIV/AIDS that appeared in Malayalam magazines and newspapers indicates that science-related cartooning is not completely absent in the regional press (Sreekumar 2009).

48. See the *Kunjamman* cartoon strip by B.M. Gafoor in *Mathrubhumi* on 8 January 2001, p. 1, and 1 August 2001, p. 1. Also see the cartoons in *Deshabhimani* on 20 July 2001 (p. 4) and 24 July 2001 (p. 9). These were the only representations of the controversies available in the newspapers under study. The absence of science criticism in the genre of cartoons is a topic that deserves further attention. A good number of cartoons appear in many of the non-Malayalam newspapers that laugh at science in general and particularly at the Indian scientific establishment. Cartoons by the reputed Indian cartoonist (late) R.K. Laxman are good examples of the same. The cartoonist once quipped,

> The subject of science is not my cup of tea; but armed as I was with the cartoonist's licence to be inaccurate and irresponsible, I saw exciting possibilities. *Science Today* [the magazine in which many of his science cartoons appeared] offered an excellent platform for this venture. So I started off drawing cartoons

> for it on things scientific, or so I thought! Soon I realised that science did not
> confine itself to cosy labs alone with their physics and chemistry, as I had
> believed.... After I passed off these items quietly as belonging to the som-
> bre world of science I could not prevent the gate crashing of some glaringly
> unscientific things!.... Yet I have taken the liberty of bringing them under the
> sacred heading of science in this book because I am convinced that they are
> bound one day to be scientific certainties in the future.

See the cartoonist's introduction in Laxman (1982). It is interesting to see how
the cartoonist developed the possibilities available within the genre to challenge
the boundaries of science by offering a powerful criticism through his drawings.

49. Personal interview with Dr C.R. Soman, 26 April 2006.

50. Personal interviews conducted with the scientists (names withheld)
on 27 April 2006. However, some other scientists from the institute suggested
that *Mathrubhumi* journalists maintained a cordial relationship with them.
Personal interviews with Dr V. Sasikumar, 27 April 2006, and John Mathai,
4 May 2006.

51. See Latour (1987, 1988, 1996) for a demonstration of the process of
translation from the purview of the actor-network theory (ANT). For the use
of ANT to analyse media reporting of science, see Neresini (2000) and Plesner
(2010).

52. Personal interview with Dr C.R. Soman, 26 April 2006.

53. Personal interview with Dr C.R. Soman, 26 April 2006.

54. Personal interview with Dr C.P. Rajendran, 27 April 2006. Journalists
also had to juggle their multiple identities and professional commitments. See
Pillai (2012: 41–2) for a discussion on how regionalism affected journalistic
professionalism during the Mullaperiyar dam controversy.

55. Press conference is a theme that is rarely studied (Bhatia 2006: 173).
A press conference has its own autonomy as a deliberative space. Analysis of
the use of language, discursive strategies and so on, in the press conferences is
important in understanding the constitution of the space (Bhatia 2006). For a
brief analysis of the press conference of two electrochemists from the University
of Utah, USA, that triggered the cold fusion controversy in 1989, see Simon
(2001).

56. Mostly these sources consisted of self-styled experts as well as scientists
who work from small, less-known, and less-established scientific institutions
and research areas.

57. Many of the journalists whom I interviewed agreed upon their use of the
internet as a source of information.

58. Personal interview with Dr V. Sasi Kumar, 27 April 2006. During my
interviews, many other scientists also made explicit their use of internet as a
source of information.

59. See Chapter 6 for a detailed discussion on the endeavour of Dr Godfrey Louis and Dr Santhosh Kumar to attain scientific acceptance and public appreciation for their research. The researchers pointed out that the preliminary debate on the alien cell theory was hosted by websites like *World Science* (http://www.world-science.net/). Personal interview with Dr Louis and Dr Kumar, 23 June 2006.

60. Some of the reputed scientists whom I interviewed spoke critically of public attempts to achieve the status of experts by using the internet as a source of scientific information.

61. Personal interview with B.R.P. Bhaskar (media analyst and journalist), 12 June 2006. *The Hindu*, for example, has a well-organized science beat. A science beat consists of a team of journalists following the developments in different fields within science and producing science news. The newspaper may have an editor or an editorial team who organize the science content. The 'Science and Technology' page in *The Hindu* (used to be published on every Thursday, and of late on Mondays), and the 'Knowledge' page of *The Indian Express* (published on Sundays), for example, present new developments in contemporary science.

62. Personal interviews with P.V. Murukan (*Kerala Kaumudi*, 12 June 2006) and Sasidharan Mangathil (*Mathrubhumi*, 26 June 2006).

63. Personal interviews with P.V. Murukan (*Kerala Kaumudi*, 12 June 2006) and Sasidharan Mangathil (*Mathrubhumi*, 26 June 2006).

64. For instance, Sasidharan Mangathil of *Mathrubhumi* is a postgraduate in geology, and he was previously a research associate in a scientific project of the CESS. He has won several awards for reporting scientific and environmental issues, among which the 'Science Journalism Award' of the Kerala State Council for Science, Technology and Environment (KSCSTE) in 2013 for news reports on landslides in Kerala deserves special mention. P.K. Prakash, who actively reported the earthquake–well collapses controversy for *Madhyamam*, is a well-known journalist who has bagged many awards for investigative journalism on issues related to science, development, and human rights. P.V. Murukan of *Kerala Kaumudi* cited the names of some of the science reporters of his newspaper and pointed out that they had at least graduate training in science. Personal interviews with P.V. Murukan (*Kerala Kaumudi*, 12 June 2006) and Sasidharan Mangathil (*Mathrubhumi*, 26 June 2006).

65. Personal interview with Sasidharan Mangathil, 26 June 2006. He opined that it was mainly graduates in Malayalam literature who used to opt for journalism. As mentioned, this trend changed in the 1990s, and the field of journalism became open to science graduates as well. I was told that there were four journalists in *Mathrubhumi* assigned to follow developments in various fields of science on a regular basis.

66. Personal interview with Sasidharan Mangathil, 26 June 2006.

67. This phenomenon resonated with the 'Matthew effect' in science proposed by Robert K. Merton, although it was media, not the internal social dynamics of the scientific community that granted them recognition. See Merton (1968).

68. See Chapter 5 of this book.

69. Personal interview with Sasidharan Mangathil, 26 June 2006. John Mathai was a scientist at the Department of Geology, CESS.

70. Personal interview with Dr V. Sasikumar, 27 April 2006.

71. Personal interviews with John Mathai, 4 May 2006, and Dr V. Sasikumar, 27 April 2006.

72. Emphasis added.

73. Personal interviews with G. Sankar, 4 May 2006, and Dr V. Sasikumar, 27 April 2006.

Conclusion

8 Science, Media, and the Question of Democracy

Whenever I explained my research topic to someone in Kerala during casual conversations, I invariably received a grim look followed by more or less the same comment: 'No Malayalam newspaper reports science as seriously, adequately, and professionally as *The Hindu* does.'[1] This baffled me initially. Nevertheless, when explained further that the study was based on the public controversies on the clinical trials, the tremors and well collapses, and the coloured rains, the preliminary scepticism quickly gave way to animated recollection of the events associated with the disputes. All the respondents were cognizant of the controversies, and identified the Malayalam dailies as their major source of information about the same. Many of them shared their personal views on the issues debated, asking for my opinion. The controversies were indubitably understood as 'science-related'. That is, the controversies were identified as involving scientists and scientific institutions. At the same time, the respondents never framed the controversies as related to 'modern science' because they considered science to be a monolithic, socially disembodied, and progressive enterprise; something sacrosanct, which had nothing to do with the regional scientists/scientific institutions and their mundane interactions with the public!

Evidently, this personal experience of mine indicates the existence of a strong social memory in the region related to the cases discussed in this book. And the role of the regional press in the creation of the same

was crucial. Our contemporary social reality is deeply shaped and repro-
duced by modern science and mass media. As a result of this, the public
critically engages with both the institutions, whereby it prepares the
ground for the emergence of an active subpolitics manifested through
the scientific public sphere.[2] Paradoxically, our general understanding
of how the public takes cognizance of science predominantly conforms
to the diffusionist/deficit model. From this latter viewpoint, the media
is perceived as distorting science, or as possessing a poor understanding
of science. Our quotidian social interactions with science contradict our
own philosophical understanding of 'modern science'. Unfortunately,
the hegemony of the latter discourse often discourages scholars working on India from examining the everyday sites of public engagement
with science and the influence of this 'everydayness of science' on
democratization of science and society.

The present study was an attempt in this direction to understand how
media facilitates an everyday politics around manufacture of risks that
encourages democratization of science. How do the citizen-public in
Kerala engage with and understand contemporary science in their daily
lives, defying the common perception (both popular and academic)
that the public is 'homogeneous', 'passive', 'ignorant', and 'unruly' in
their contact with science? What role does the regional media play in
this process? What is the relationship that exists between science, media,
and politics in Kerala? Why are several controversies over science mani-
festing through the regional press? How do public controversies over
science staged by the Malayalam newspapers contribute to deliberative
democracy? What does the shaping of a scientific-citizen public tell us
about science in liberal democracies? These were some of the pertinent
questions the book tried to engage with.

As the present study has demonstrated through the analysis of the
risk controversies, the frequent eruption of public deliberations is
indicative of the emergence of a strong subpolitics around science in
Kerala. The subpolitics of science is deeply linked to the wider democratization processes. Changes in our imagination of democracy are
always reflected in the emergence of new ways of understanding and
engaging with science.[3] Or, in other words, scientific knowledge and
social orders are co-produced (Jasanoff 2004). Hence an examination
of the scientific public sphere is important to understand the democratization process in contemporary societies.

Revisiting the Controversies

A comparative assessment of the three scientific controversies under study is attempted here (see Table 8.1). This helps us understand their similarities and differences and the specificities of the deliberative process in each case. The comparative analysis will also bring into focus some of the unique features of the civic epistemology of Kerala

Table 8.1 Comparison between the Controversies

	RCC Clinical Trials Controversy	Earthquake Controversy	Coloured Rain Controversy
Causes	Questions of research ethics and patients' rights raised by a group of scientists about the clinical trials on cancer patients	A series of unknown and unusual phenomena and the emergent risks and uncertainties	Unknown and unusual phenomenon and the public discontent about the official scientific explanation provided
Risk Perception	Subjection of patients' body to unethical medical experiments; threats to treatment-seeking practices	Risk of a looming natural disaster; possible failure of the Mullaperiyar dam due to increased seismicity	Coloured rain as indicative of the manifestation of unknown environmental problems; possible connection with an impending natural disaster
'Science' Addressed	Contract research organizations and clinical trials; regulation of drug research in India	Complexity of the phenomena; contradictions in scientific interpretations; old and new technologies	Contradictions in scientific interpretations; debate among various research groups; the theory of Panspermia

(Cont'd)

Table 8.1   *(Cont'd)*

	RCC Clinical Trials Controversy	Earthquake Controversy	Coloured Rain Controversy
Scientific Disciplines/ Fields Involved	Cancer research (oncology)/drug research/medical sciences in general	Earth sciences, hydrology, geology, ocean science, atmospheric science, environmental sciences, and dam engineering	Atmospheric science, biological science, environmental sciences, astrobiology
'Experts'	Scientists from a single institution	Several scientists and engineers from different institutional and disciplinary backgrounds; large number of self-claimed experts/scientists	A few scientists from a limited number of institutional and disciplinary backgrounds; a small number of self-claimed experts/scientists
Research Networks and Collaborations	International collaboration and networking as triggering off the dispute; public scepticism about research collaboration with an American university	Regional and national scientific institutions collaborated to resolve the scientific disputes involved	International collaboration and intellectual exchange was fruitful for both sides
Professional Relationship between Scientists and Journalists	Scientists actively seeking journalistic support to win the controversy	Journalists contacting certain scientists and institutions for exclusive information; wide networking between the two	Wide networking between the two professional groups

(Cont'd)

Table 8.1 *(Cont'd)*

	RCC Clinical Trials Controversy	Earthquake Controversy	Coloured Rain Controversy
		professional groups; assistance offered by media establishments to researchers in collecting data	
The Regional Media	Visible involvement as agenda broker and gatekeeper; active role in maintaining the controversy	Visible involvement as a spokesperson for the public; bringing out the complexity and contradictions of science to the public	Lower visibility as an agenda broker; opening up a public channel for communication among experts
The Scientific-Citizen Public	Defending a regional institute while critiquing unethical practices in science; argumentative; actively seeking participation in the democratic governance of science	Argumentative and sceptical; confrontational; anxious about the risks and uncertainties; claiming lay-expertise	Curious spectators; less argumentative
Role of the Government	Actively defending science/research institution; role as adjudicator in the dispute	Actively seeking expert advice to resolve social/political crises	Mostly absent
The Reach/ Scope of the Controversy	Regional, but deeply connected to global concerns; widely reported	Mostly limited to the region	Regional and international; widely reported internationally

(Cont'd)

Table 8.1   *(Cont'd)*

	RCC Clinical Trials Controversy	Earthquake Controversy	Coloured Rain Controversy
	in international media; great policy impact in India		
Involvement of Civil Society Organizations and Movements	Active participation	Mostly absent	Totally absent
Credibility of Scientific Institutions and Trust in Experts	Credibility lost and regained; reassurance of public trust in one group of researchers	Gradual erosion of trust in experts; credibility of many institutions compromised; national institutions and their experts enjoyed greater public legitimacy	Moderate loss of trust and credibility
Closure	Complete closure	Abrupt ending after attaining partial closure at earlier stages	Partial closure

expressed in the scientific public sphere. The discussion on deliberative democracy and the scientific-citizen public towards the end of the chapter is premised on the comparison of the public controversies made in this section.

The clinical trials controversy erupted when a group of scientists attempted to defy the professional hierarchy within their research institute, raising questions of research ethics and patients' rights in the media. Therefore, only a very small number of experts were involved in the deliberation and the main protagonists were the researchers from the same institution and their allies. There were no self-claimed experts involved unlike in the other two controversies, where a large number of participants claimed 'scientific expertise' and identified themselves as 'scientists'. There was strong risk perception regarding unethical practices in medical institutions and their impact on patients' bodies. Intrusion of commercial interests in drug research was seen as part of an emergent global politics that creates serious threats to the treatment-seeking behaviour of the public. The main causes for the controversies on micro-earthquakes and coloured rains in Kerala were the incongruities in the regional scientists' interpretation of the unusual natural phenomena. The absence of well-established and legitimate ways of interpreting geological phenomena resulted in the eruption of a series of disputes and consequential activation of the scientific public sphere. The possibility for a big earthquake in Kerala in the near future because of the changes in the region's tectonic structure was perceived as a risk in the context of the manifestation of unnatural geological phenomena such as well collapses and continuing micro-tremors. The seismic activation of the region also gave rise to public anxiety about the possibility of dam failure at Mullaperiyar. The structural vulnerability of this old dam was perceived as accentuating the risk. In the years that followed, the dam continued to trigger public angst about its collapse during heavy monsoons. The coloured rain controversy was more 'manageable' as there was only a single unusual phenomenon that was the cause of public anxiety. The public perceived the coloured rains as a symptom of unknown environmental problems or changes in the ecosystem. The coloured rains were also perceived as linked to earthquakes, along with well collapses and other unnatural phenomena. During the coloured rain controversy, deliberations in the scientific public sphere was more restricted: a limited number of experts from a couple of scientific institutions and disciplinary backgrounds participated in the controversy. While the self-claimed experts and scientists from national institutions received more media attention in the former case, the coloured rain controversy witnessed a single group of researchers gradually claiming public legitimacy with their dexterous use of the mass media.

The deliberations in the scientific public sphere strongly suggest that the idea of science did not conform to its positivist construal as homogenous and universal. Science was always understood in its heterogeneity and complexities. The deliberations incessantly

pointed towards its particularities, although some actors (especially
those who were threatened with a loss of credibility) unsuccessfully
tried to defend themselves by invoking the positivist image of science
as a rhetorical strategy. Public criticism was always focused on some
specific group of experts or on a particular scientific institution doing
research in a particular field or was directed towards a specific piece of
expert advice, experiment, or interpretation. Loss of trust in a certain
group of experts always led to a shift of trust to another expert group
or scientific institution. As Steven Shapin (1994: 19) suggests, distrust
is always localized, and grounded on a system of trusting—'[d]istrust
is something which takes place on the *margins* of trusting systems'.[4]
From this point of view, the public's loss of trust in experts and erosion
of institutional credibility were situated events embedded in a deeper
system of public trust in science as a reliable system of knowledge. In
the RCC controversy, the changing dynamics of drug research was the
focus. The deliberations centred on its growing commercialization and
the role of contract research organizations (CROs). The RCC and its
clinical trials particularly came under public scrutiny. And there were
two specific groups of researchers at the centre of the controversy. In the
earthquake–unnatural phenomena controversy, the focus was on the
Centre for Earth Science Studies (CESS) in Thiruvananthapuram and
some other regional and national institutes. Specific scientific explanations and theories were discussed and certain experts were criticized.
Also, the Mullaperiyar dam and its structural weakness were contrasted
with the technical resilience of the Idukki dam. The coloured rain controversy witnessed public debate on the CESS scientists' interpretation
of an unnatural phenomenon that was challenged by other groups of
researchers. The theory of Panspermia became the focus at a later stage,
and the collaboration between a small team of researchers from Kerala
and the researchers at the Cardiff Centre for Astrobiology, UK, became
the centre of attention.

The major scientific disciplines and areas of expertise involved varied
from case to case. In the clinical trials controversy, cancer drug research
in a regional institute was the focal point. Slowly the global political
dynamics of biomedical research also came under the purview of the
debate. Sporadic manifestation of a wide range of strange geophysical
phenomena in the earthquake controversy initiated the involvement
of several areas of expertise in the deliberation that followed. In the

coloured rain controversy researchers from multiple disciplinary back-grounds (especially the life sciences and atmospheric science) asserted exclusive interpretative claims over the phenomenon.

Scientific expertise of an actor was accepted or rejected in the scien-tific public sphere based on several factors. The clinical trials contro-versy witnessed the retrieval of the lost public credibility of the institute through skilful manoeuvring in the scientific public sphere, whereas the earthquake controversy demonstrated a continuous erosion of trust and credibility. During the disputes over unusual geological phenom-ena such as well collapses and coloured rains, it was often the case that scientists from national institutes were conferred with greater episte-mological authority, and scientists from the regional institutes were rel-egated to a lower status in the public debate. The 'scientific' arguments of the self-claimed experts helped in this process, but they themselves never enjoyed much public legitimacy. The cases of coloured rain and clinical trials revealed that it was those researchers who made strong coalitions with other actors such as the journalists became triumphant in gaining public acceptance. The cases show that the category of experts was not fixed and rigid in the public domain, but evolved in the course of deliberation. The expertise of the actors were differentially employed and negotiated in each controversy.

The nature of intellectual networking of the scientists who were involved in the controversies along with their allies at regional, national, and international levels also varied from case to case. In the second and third controversies these intellectual collaborations were productive in deriving legitimacy and reclaiming epistemic authority. On the other hand, the collaboration itself became controversial in the RCC controversy. The clinical trials carried out by the RCC scientists for the researchers of the Johns Hopkins University (JHU) invited public criticism, being perceived as an instance of uneven power relations within the medical research system and the consequent colonization of patients' bodies in developing countries. Researchers from the RCC occupied a hierarchically lower position in the global network of sci-ence, for they were seen to be undertaking clinical trials of drugs devel-oped by the former. In the earthquake controversy, the networking was among regional and national institutions. The collaboration largely benefited the regional scientists as they could regain trust and cred-ibility to some degree with the help of the national scientists who were higher up in the hierarchy. There was little collaboration with experts outside the country in this case. In contrast, the case of coloured rains saw a group of scientists successfully rising to fame due to their international research collaborations. Their intellectual alliance was more even in its power relations.

Professional interaction between scientists and journalists was exceptional in all the three contexts of deliberation. Both the professional communities actively sought help from each other. For the journalists, camaraderie with scientists was inevitable to seek scientific information and expert opinion. Journalists were under pressure to gather knowledge in all its diversity as their readers sought scientific information and explanations, and technical advice through the media. Skilful reporting of science was therefore inevitable, and gaining the confidence of scientists along with developing long-term professional relationships with them was crucial for journalists. For the scientists, media exposure was helpful towards gaining public recognition and fame as well as rewards within the scientific hierarchy. More public exposure to their research was expected to be translated into increased peer recognition too. Close association with journalists was quintessential to accomplish these professional goals. The cooperation between researchers and journalists also led to developing innovative practices such as scientific data collection with the help of media's infrastructure during well collapses and coloured rains. Intense professional collaboration of the communities of journalists and researchers was one of the most noteworthy features of the science–media coupling in the region.

Media's role in the controversies was case specific. All the controversies were staged by the media and they also played the part of actors in the deliberations. However, this dual role of each newspaper varied in each controversy. The earthquake controversy saw their active role in revealing the internal ambiguities and contradictions of science to the public and thus intensifying the public debate. The newspapers performed the role of experts on several occasions, actively making 'scientific' observations and reasoning against or jointly with scientists. In the RCC controversy, the regional press was deeply divided into two camps, in support of one or the other group of the RCC scientists. The regional press played a unique role in this case, exhibiting their civic responsibility by defending the institution and arguing for better regulation of scientific research. In contrast to these cases, the coloured rain controversy was marked by the absence of the Malayalam newspapers as actors in the deliberations. The regional press was mostly involved in staging the debate by actively pursuing the interpretations of the phenomenon by different research groups and engaging them in scientific exchanges by functioning as a public channel for inter-specialist communication. The scientific public sphere was an important site of production and validation of knowledge for science, as the mediated deliberations demonstrated.

The characteristics and behaviour of the scientific-citizen public varied from one controversy to another. During the RCC controversy they were highly argumentative and critical of the unethical clinical trials, but simultaneously defended the regional institute. Those who supported the rival scientists were also emphasizing the regional importance of the institution while demanding more public accountability from the researchers. In other words, they actively sought more public participation in scientific governance and argued for better regulatory mechanisms, in order to uphold the scientific esteem of the institute. The public became organized into social movements, political parties, citizen forums, and discussion groups to participate in the RCC controversy, which was a social phenomenon nearly absent in the other cases. The controversies over earthquakes and well collapses constituted a scientific-citizen public highly concerned about the risks and uncertainties generated by earthquakes and allied geophysical phenomena. They actively sought scientific explanations and information from the regional research community. The fact that the scientific community was deeply divided in their assessment of the risks heightened public anxiety. Therefore, the scientific-citizen public turned to be largely critical of the experts and their knowledge capabilities. Being highly argumentative and confrontational, they were at loggerheads with the experts in local contexts of interaction. The public asserted their lay-knowledge and generated fringe theories and alternative explanations in their attempt to counter the expert interpretations once their trust in scientists waned. The coloured rain controversy revealed that the scientific-citizen public was curious about the spectacular phenomenon and the scientific investigations that followed. They were largely silent spectators to the scientific communication between research teams and made less interventions than in the other controversies.

Governmental interventions happened only in the first two contro-versies. In the first case, the state was overtly defending the regional research institute and openly protected the researchers who conducted the clinical trials. The state government was very much an ally of the coalition led by the director of the institute, though it tried hard to proj-ect its role as a neutral adjudicator before the public by presenting itself as highly concerned about the issues and apprehensions raised. The government's initiative to appoint enquiry commissions helped the controversy attain closure. The state largely played a proactive role in the earthquake controversy too, trying to resolve it by constituting expert committees and paying attention to the public and media responses. The government often took a non-partisan stance in the controversy, even though the Mullaperiyar dam dispute was an exception. When the dispute emerged, the Government of Kerala, the scientists belonging to governmental institutions, and the regional press worked together as a team to scientifically challenge the Tamil Nadu government's demand for an increase of water level. The public in Kerala largely supported this move. Science and regional politics were deeply interwoven in an explicit manner in this controversy. In the coloured rain controversy, the state was largely absent as an adjudicator and made few interventions, although it had directed the CESS to begin a scientific probe on the phenomenon. The low intensity of public criticism and a lack of strong perception of risks seem to be the reasons for the regional government's withdrawal from the case unlike its high involvement in the others.

The clinical trials controversy and the coloured rain controversy were reported in the international media for the participation of research institutions/scientists from the West. The former gained international traction because of the involvement of the JHU. Reputed peer-reviewed journals like *Science* as well as the popular media reported the events. The controversy was internationally perceived as exposing the malprac-tices in medical research and raising ethical concerns about the neo-liberal, market-oriented transformations in biomedicine. The coloured rain phenomenon also attracted global and national media attention, thanks to the international research collaboration that emerged. In contrast, the earthquake controversy was mostly limited to the region. Although the well collapses were reported in the national media, it was never seriously deliberated outside the region. The national scientific community showed some interest in the phenomenon and many researchers participated in the deliberations and collaborated with the research projects at the regional level.

The public controversies facilitated a proliferation of popular science literature that included books, essays in magazines, and editorial page articles in newspapers. A significant venture during the earthquake controversy was the publication of a popular science book that examined different aspects of earthquakes and Kerala's geological structure, as a collaborative effort of a group of experts, which also included a journalist.[5] Many of these authors were directly involved in the deliberations in the scientific public sphere. A perusal of the articles and books produced in connection with the disputes suggests that popular science had acquired new meanings and functions during risk controversies; the stabilized knowledge normally presented in popular science literature became re-situated within a wider system of contested knowledge. They acquired extreme interpretative flexibility and became resources for the actors to participate in the deliberations, although the original thrust of these publications was to present stabilized and non-controversial science.[6] On the other hand, popular science published during the public debate over the Mullaperiyar dam could be included more aptly in the category of 'controversial popular science', which was intended to defend a standpoint in the dispute.[7] Newspapers published numerous popular science articles on the editorial page during the controversies in order to explain various aspects of the technical issues involved. These articles served both the functions of popular science, either providing resources for the deliberative public, or taking positions in the controversies.

Another interesting phenomenon was a burgeoning of popular cultural productions around these controversies. Several popular films, documentaries, and novels surfaced in the aftermath of all the three controversies, opening up popular culture as a site for negotiating anxieties about science. *Apothecary* (2014, Malayalam), a mainstream film, portrayed the ethical dilemmas of a neurosurgeon in a profit-oriented, private hospital in performing clinical trials on his patients. While this film did not make any direct reference to the RCC clinical trials controversy, there were several others with explicit references to the other two controversies. Rahul Roy's science fiction disaster film, *Dam 999* (2011, English), that visualized the failure of an old dam created much public and legal dispute in connection with the Mullaperiyar dam controversy. Tamil Nadu banned the film alleging that it was propagandist. The Supreme Court rejected the director's petition that challenged the ban and observed that the film's release unnecessarily aggravated the fears and anxieties of the public.[8] It was, nevertheless, released nationwide. Rahul Roy also produced a documentary film (*Dams: The Lethal Water Bombs*, 2011) and a novel (Roy 2011) on the threat posed by dams. In response to the feature film, officials of the Tamil Nadu Public Works Department produced a documentary film titled *The Mullai Periyar Dam: Problems and Solution* (details not available) to present Tamil Nadu's perspective.[9] *Lingaa*, a recent Tamil movie (2014, dir. K.S. Ravikumar) with superstar Rajinikanth in a double role, narrated the politics around an old dam and the protagonist's (son of the raja of the princely state who built the dam during colonial times) endeavours to stop the attempts of a corrupt local politician to demolish it. There was also a novel published in Malayalam based on the Mullaperiyar dam dispute titled *Jalabhramangalil Njan* (Siddeek 2010). The BBC produced a documentary film in 2006 on the coloured rain phenomenon in Kerala and highlighted the alien cell theory (*We Are the Aliens*). *Red Rain* (2013, dir. Rahul Sadasivan), a science fiction thriller movie in Malayalam, has its storyline strongly influenced by the unnatural phenomenon. These cultural productions require close analysis which is beyond the scope of this book. However, it can be stated that such an intense proliferation of cultural texts clearly points to the deep cultural impact of these controversies and the anxieties and uncertainties generated.

As mentioned earlier, there was a strong involvement of citizen forums, political parties, youth organizations, and employee's associations in the clinical trials controversy. Public demonstrations and debates were organized by them as parallel sites of deliberations that pushed certain concerns strongly into the media-generated scientific public sphere. The presence of these mini-public spheres radically enhanced the political impact of the controversy, and demanded responsible intervention from the government. Contrastingly, the involvement of civil society organizations was minimal in the earthquake controversy and totally absent in the deliberations on coloured rains.

The credibility of science and the public trust reposed on scientists took different turns in different contexts. While the scientific community of the region experienced a massive erosion of public trust in the second and third cases, researchers from the RCC had to face only a temporary and partial loss of trust and credibility. The decline of public credibility of the institution was successfully curbed through deliberative moves and the trust in scientists was eventually restored. In contrast to this, in the earthquake controversy, the trust and credibility were never regained completely, and researchers from certain governmental institutions like the CESS faced the loss of trust of the public more than other experts. The coloured rain controversy started with a strong opposition from the regional scientific community against the research findings of the team of experts from the CESS, and the rise of alternative hypotheses was catalysed by the institution's already diminished public credibility, thanks to the earthquake controversy. The institute and its scientists suffered a further loss of public trust during the coloured rain controversy when their findings were strongly challenged in the scientific public sphere. At the same time, the endeavour of a small team of researchers to achieve international academic recognition (which they enjoyed to some extent) with the help of media did not find much support from within the region.

There were marked differences in the way closure was attained in each case. The RCC case witnessed a complete closure of the dispute with apparent public consensus on the need to safeguard the prestigious public institution of cancer research. Most of the key regulatory issues under deliberation were resolved by the intervention of governmental regulatory bodies. The controversy had a deep policy impact on the regulation of drug research in India. While the earthquake controversy ended up in an abrupt and open finale without any consensus about the phenomena or the questions under debate, there was partial closure to the coloured rain controversy. A shift in the media attention towards the terrorist attack on the World Trade Center in New York on 11 September 2001 was the major reason behind the sudden demise of the controversy over the earthquakes and other strange, unnatural phenomena in Kerala. Based on the new scientific assessments in the aftermath of the strong expression of public perception of the risk of earthquakes in the region, Kerala was promoted from the second (low intensity/low damage risk) to the third zone (severe intensity/moderate damage risk) in the seismic vulnerability map of the Indian peninsula.

Regional Science and Risk Regulation

The comparative analysis of the cases also offers insights into some of the significant characteristics of the scientific establishment of the region.[10] As the cases we have examined clearly demonstrated, the regional scientific institutions and their researchers were under rigorous public scrutiny. Several regional and national research institutions came into public attention during the scientific controversies. It is evident from the controversies that most of the scientific institutes which took part in the deliberations were autonomous research institutions under the patronage of the state. Some of them were under the state government while others were affiliated to the Government of India (GoI).[11] The research wings of different government departments like the Kerala State Electricity Board (KSEB), the Kerala State Groundwater Department (KSGD), and the Department of Mining and Geology (DMG), Government of Kerala, also participated in the deliberations. Being under the patronage of the state, these institutions were the nodal agencies of regulatory science and were expected to provide expert advice. The government, politicians, and policy makers relied heavily on these institutions for standard scientific explanations and recommendations towards effective risk governance. Many national research institutes situated outside Kerala also were actively involved in the controversies and provided technical advice when the regional institutions faced credibility crises.[12] These institutes and their researchers were drawn into the regional controversies to offer more credible and legitimate expert opinions, for the public credited them with greater epistemic authority than the regional scientists. In some cases, their interventions brought in closure to the dispute.

Research departments of various universities in the region also participated in the deliberations. The scientists from two research departments of the Mahatma Gandhi University, Kottayam, were actively involved in the coloured rain controversy.[13] The research facilities at the Cochin University of Science and Technology (CUSAT)[14] as well as the Sree Chitra Tirunal Institute for Medical Sciences and Technology, Thiruvananthapuram,[15] were utilized by scientists from less privileged institutions in their investigations and this helped the weaker institutions to acquire peer recognition and public acclaim. The physics department of the Newman College at Thodupuzha conducted its own independent investigations on the coloured rain phenomenon. It also attempted to continue research with the support of the photonics department of the CUSAT. There were instances of professors and teachers contributing to the public debate in their individual capacities too. These researchers could not attract much attention, apparently due to their low academic capital and the consequent lack of epistemic authority.

Several regulatory bodies and funding agencies of the central government appeared as stakeholders in the controversies, responding to public criticism. The Department of Science and Technology (DST) of the GoI, the Indian Council of Medical Research (ICMR), and the Drug Controller General of India (DCGI) were key arbitrators in the controversies. These governmental institutions intervened in the scientific disputes by constituting new regulatory/ethical committees assigned to reformulate policies and to set fresh guidelines. The intervention of governmental regulatory agencies aimed at resolving the disputes and insulating science from public scrutiny. For instance, the RCC clinical trials dispute achieved closure predominantly due to the timely regulatory interventions of the ICMR and the DCGI. However, many of these regulatory bodies themselves were targeted for their inability to fix the standards of scientific research. Public resentment regarding poor formulation of research guidelines by the ICMR, and the role of the DCGI and the RCC ethical committee in the clinical trials controversy exemplify this.

Besides this, the state and central governments constituted several enquiry commissions and advisory committees at different stages of the deliberations in order to resolve the tensions and to attain closure. These expert bodies, constituted in the context of particular disputes, quite often lacked public trust and confidence and came under severe public criticism. The Parikh Commission constituted by the state government to end the RCC clinical trials controversy is a good example. The case of unusual geological phenomena witnessed the appointment of several expert committees to investigate the issues under dispute, and their findings often initiated fresh controversies rather than ending the existing ones. An exception to this common trend was the expert committees appointed by the Kerala state government to examine the vulnerability of the Mullaperiyar dam which enjoyed great public acceptance, although at the national level their advice was not always given much credence.

For many of the scientific institutions who participated in the controversies the technical questions involved were different from their in-house research projects. The geological structure of the region and the increasing seismicity in Kerala were not primary research concerns for a number of experts who participated in the controversy over recurrent micro-earthquakes. Two renowned scientists, who had actively represented the CESS in the dispute over tremors and well collapses, revealed that their contemporary research was on the geological structure of the Andaman and Nicobar Islands rather than Kerala, since the latter was not a seismically active zone and therefore scientifically less interesting.[16] Many researchers were introduced to the scientific issues related to the controversy only when the research institutions they were affiliated to or the government demanded technical reports or when they were nominated to expert committees. Therefore, in several instances the experts did not have proper knowledge of the subject or possessed only limited technical expertise to investigate the problems. Also, the time constraints prevented them from developing domain knowledge on the topic under investigation. Their interest in the subject eroded when the controversies disappeared, and thereafter they reverted to their ongoing research projects. Only a few scientists took further interest in designing new research projects related to the controversial issues.[17]

Ironically, the general disinterestedness of the researchers was exacerbated by the novelty and complexity of the natural phenomena. Most of the geological incidents discussed were completely new to the regional scientists and the phenomena were not amenable to the concerns and methods of any single scientific discipline. This could have led to multidisciplinary investigations, but the researchers who attempted to investigate the phenomena mostly tried to understand them from within the scope of the disciplines they were trained in. The experts' lack of persistent enthusiasm for the issues and their incompetence in innovatively framing the problems prevented them from arriving at satisfactory conclusions. Experts were often aware of the need for interdisciplinary and multidisciplinary approaches, but failed to embark on them.[18] The CESS had identified this grave shortcoming and convened a meeting of its scientists to collectively develop an interdisciplinary framework to investigate the geophysical phenomena but could not proceed further.[19] Professional rivalry between scientists within the institute appeared to have hindered their collaboration, and disciplinary and institutional rigidities further sabotaged multidisciplinary partnerships. The same could have been the major reason behind the failure of many inter-institutional expert committees constituted by the government.

The absence of a formal, institutionalized regulatory mechanism for assessing and mitigating risks through rigorous engagement between experts and citizens became exposed during the risk controversies. The scientific research system in Kerala was totally unprepared to take up this additional role of providing technical interpretations and expert advice during the sudden manifestations of risks. The expert committees constituted under the direction of the government tried to perform the task of risk assessment, but they were not part of a well-established regulatory system.[20] Similarly, the RCC controversy revealed the weakness of institutional regulatory mechanisms within the regional research system. Better regulative measures for ethical conduct of drug research were identified as an urgent requirement by the public and the government at this juncture. Seemingly the general vulnerability of the regional research system in Kerala got reflected in the shortcomings of the regulatory practices.[21] The activation of the scientific public sphere was hence crucial for exposing these limitations and prodding the government and the scientific establishment to develop a better institutional mechanism for both the democratic governance of science and regulation of risks. The scientific public sphere functioned as an informal mechanism that efficiently enabled public participation in risk regulation and S&T governance.

The fact that the scientific public sphere came into existence in the context of strong risk perception nevertheless moulded it further into a site for public engagement with science-in-the-making. It actively functioned as a public site of knowledge production, complementing the laboratory/experimental field by bringing together a large spectrum of actors who intensely engaged with science. The scientific public sphere often created channels for technical communication between researchers, and constituted an 'extended peer-group' of lay-public who reviewed the scientific claims and interpretations offered to them.[22] It functioned as a trade zone between different epistemic cultures within science, and opened up dialogic spaces between scientific and popular cultures. Beyond technocratic regulation of science, the scientific public sphere facilitated a re-politicization of science that thwarted the conventional relationship between science and politics. However, it was not a space that allowed anything and everything in the name of deliberation. The performances of the actors on the front stage of the theatrical structure of the scientific public sphere were dependent on the backstage management. There was a strong filtering mechanism being employed in the staging and regulation of public controversies. Each controversy developed its own basic civic epistemological frames that oriented the deliberations in a particular direction.

The civic epistemology—'culturally unique, historically shaped and politically grounded, tacit and often institutionalised "collective knowledge-ways"' according to Sheila Jasanoff (2005: 255)—reflected in the public deliberations was predominantly *contentious*;[23] scientific explanations and expert advice were actively sought, but rigorously criticized and compared with alternative propositions. At the same time there was a deep reflexivity about the very functioning of the scientific public sphere. The need for public deliberations in democratizing science was incessantly ascertained and every attempt to shield science from social audit was challenged. Scientific public sphere presented science as existing through its particularities such as expertise, theories and interpretations, institutions, regulatory protocols, data, research collaborations, instruments, diagrams, disciplines and so on, and defied the positivist imagination of science as a monolithic entity. Public deliberations focused on the social embeddedness of science and resisted ideological attempts to separate out truth from power, knowledge from politics, and the natural from the social. Risk issues were hence framed as 'techno-social' in the scientific public sphere. The power of science was not readily accepted in thinking about problems, and the experts had to be attentive to lay-knowledge, public anxieties, and risk perceptions to be successful. Those who subscribed to the old paradigm of risk management utterly failed and their credibility got compromised in public.

The emergence of a scientific public sphere marks a shift within the civic epistemology in the 1990s from an earlier mode of public engagement with science. The public participation earlier had been shaped by the adjudicating role of the Kerala Sasthra Sahithya Parishad (KSSP). The regional press had largely restricted itself to reporting debates on science by privileging the perspective of the KSSP. The movement constructed rational frameworks for risk assessment and technological governance, and spoke for the wider public interest. Civic epistemology of the period was consensus-seeking, as the deliberations were directed towards reaching the most scientifically grounded solution to the problem at hand. Wider and more inclusive deliberations were bracketed out, since the KSSP perceived its role as *representing* the wider public. Scientific institutions and experts were criticized for their narrow perspectives informed by scientism, and the KSSP represented the possibilities of developing a people-oriented science which was broader in its outlook in solving technical problems. The shift of the site of public engagement with science from the KSSP to the mass media in the 1990s radically altered the civic epistemology into a contentious mode when the scientific public sphere emerged as an informal institutional mechanism for rigorous and open-ended public deliberations on risks. The debates were never oriented towards reaching scientifically grounded agreement on technical governance, although sometimes tentative consensus was achieved in the course of deliberation. No single actor or social group could control the debates. Risks were constructed through deliberations and a spectrum of possible interpretations and solutions were mooted for the citizens, local communities, and mini-publics to reach their own logical conclusions towards effective regulation of their social worlds. Civic epistemological frames that enabled the filtering mechanism emerged during the deliberations—these frames were not prefixed unlike the previous phase. However, the contentions in the scientific public sphere were still 'scientific' in a novel sense: ideas and interpretations that were not articulated in a 'scientific language' were discouraged, and those who did not speak the scientific language were excluded from the scientific-citizen public.[24] This language was but more inclusively shaped than the technical language of scientific experts, to hold a wide spectrum of enunciations.

Multiple Publics of Science

Different forms and categories of the scientific-citizen public were constituted in relation to the scientific public sphere. The wider public (that is, the reader-public) was in itself mass mediated, for it was the scientific public sphere that constituted them as audience. Also, the reader-public represented themselves in individual and collective capacities as actors by intervening in the deliberations on the front stage. They also materialized into mini-publics in local contexts of engagement, but their constitution and visibility were dependant on their mediation as actors in the scientific public sphere, as we have observed in the previous chapters. The scientific-citizen public generated by mass media in Kerala and manifested through the scientific public sphere in the new millennium were argumentative, knowledgeable, and reflexive about their engagement with science and the democratic process they were participating in. They were a newspaper-reading public that co-evolved with the political culture of the region. The subpolitics of science facilitated by them found its expression in the scientific public sphere.

Through the engagements of the scientific-citizen public, politics re-entered science and the technical became a site of pluralistic politics. The emergence of the scientific-citizen public who offered a critique of science was agential in facilitating this new democratic vision. The scientific-citizen public can be situated against the older, 'one-dimensional forms of citizenship' (Irwin and Michael 2003: 138). According to Leach, Scoones, and Wynne (2005), the mainstream approaches to the question of citizens' engagement with science revolve around this definition of one-dimensional citizenship rooted in the liberal political theory. Citizens in liberal democracies are 'either expected to engage passively with expert scientific institutions, especially those linked to the state, or to participate in forums orchestrated by such institutions' (Leach, Scoones, and Wynne 2005: 12). This innocuous form of citizenship (that is more in conformity with the deficit model of public understanding of science [PUS]) is contrasted with

> a [new] model of the citizen as a more autonomous creator and bearer of Knowledges located in particular practices, subjectivities and identities, who engages in more active ways with the political institutions of science ... who do not act solely as individuals, as in liberal theory, but through emergent, and sometimes global, social solidarities that may unite people around particular issues and visions, whether these be fluid and shifting with circumstances, or more lasting. (Leach, Scoones, and Wynne 2005: 12)

The scientific-citizen public of Kerala, from this vantage point, is not a homogeneous group of citizens organized under the patronage of science, media, or the state, but the politico-epistemological embodiment of a heterogeneous network of citizens, knowledge, and resources made possible through the subsystems of science, mass media, and politics which become more and more flexible and complex in their form and functioning. This new form of scientific citizenship started evolving in Kerala in the late 1990s by differentiating itself from an older form of public engagement with science shaped by the KSSP, the people's science movement. Transformations in the structure of regional media in tandem with greater manifestations of modernization risks made possible a new politics that has the imperative of democratization of science at its core.

In more recent times, it seems that public engagement with science in Kerala has further shifted its forms and sites, reshaping the scientific public sphere. Public deliberations over the exploitation of ground water by a Coca Cola plant in Plachimada (Palakkad district, 2002), the controversial spraying of endosulfan, a lethal pesticide, in the cashew plantations of Kasaragod district (2002 onwards), and a spate of controversies over industrial pollution and developmental projects have emerged since then.[25] There is a mushrooming of controversies over waste management. Similarly, public deliberations over high pesticide content in the vegetables or about the Western Ghats Ecology Expert Panel (WGEEP)[26] open new windows to the recent transformations of the scientific public sphere. While mass media continues to shape these engagements, encounters outside the scientific public sphere as well as conflicts in risk perceptions have become more visible in these controversies. A single public controversy can now increasingly harbour multiple publics and their variegated engagements with science. These multiple publics and their conflicting voices are gradually getting recognized in media deliberations.[27]

The deliberative mode of public engagement with science manifested through the scientific-citizen public in Kerala is but acutely constrained by its own democratic imagination. The scientific-citizen public co-evolved with the subpolitics of science around the manifestation of risks. The discursive construction of risks was simultaneously the crucible for the production of the scientific-citizen public. In other words, the social definition of risk is radically linked to and shaped by the politics of engagement; the imagination of democratization of science is, hence, structured through the risk perception of the scientific-citizen public. Therefore, the process of construction of the scientific-citizen public eclipsed alternative modes of engagement with science based on radically different perceptions of risks.

For instance, dalit and women political subjects were largely absent as knowledgeable participants in the scientific public sphere—this was because of the constraints of the democratic imagination to be open to their varied experiences and perceptions of risk. As pointed out elsewhere, dalits and women were represented by other actors, and their concerns became (mis)translated to conform to the risk perception of the scientific-citizen public. The RCC controversy is a case in point: the dalit/women patients' momentary appearances in the scientific public sphere were through their representation by actors in the rival coalition as victims of the unethical drug experiments. During the well collapses, women were visually represented as silent onlookers and/or victims of the unnatural phenomena. Their representation as victims was premised upon a deeper politics that denied them politico-epistemological agency to constitute/be part of the scientific-citizen public. The fact that the decades of the origin and evolution of the scientific-citizen public in Kerala were also the period of strong articulation of a new dalit and feminist politics further exposes the limitations of this new democratic imagination.[28]

What could be the alternative modes of engagement with science by these potential publics? Answering this question demands fresh explorations into Kerala's social history of science, although one event that offers fresh possibilities for theorization may be shared. During the mass sterilization camp convened in Ernakulam in 1971 as part of the Family Planning Programme (FPP), the women who underwent the sterilization procedure collectively asked for compensation from the welfare state 'for their "sacrifices" for the nation' (Devika 2008: 174). The women who raised this demand challenged the Nehruvian developmentalist state's demand of unconditional sacrifice from the population groups by reconstituting themselves as political public (see Varughese 2015b). Devika (2008) also quotes a family planning worker's complaint about the unruly behaviour of women:

> During festivals like Onam, Christmas and other special occasions, women commonly approach FP [family planning] centres more frequently to insert and remove IUDs [intrauterine devices]. This is because they get paid for insertion. They claim an incentive from one FP centre and get a loop inserted. When in need of money again, they put up false complaints like stomach pain and bleeding to get it removed. Then they go to another FP centre, get an IUD inserted, claiming an incentive.[29]

Such novel political encounter with the technoscientific project (FPP) articulated by these women (whose seemingly 'lower' caste identity was obfuscated in the FP worker's letter to the newspaper) was mediated through their technologically altered biological bodies. Such instances of citizenship claims based on biology, which Adriana Petryna (2002) calls 'biological citizenship' in another context, are constitutive moments of a radically different kind of scientific public. This public's perception of and engagement with risks is characteristically different from that of the scientific-citizen public; technoscientific risks here become livelihood options and hence 'a life-affirming resource' for these *quasi-publics* (Varughese 2012: 248). Quasi-publics, who expose the very limits of deliberative democracy that nurtures the dream of inclusiveness while paradoxically upholding the political agency of scientific-citizen publics, thus pose serious challenges to the scholarly debate on public regulation of technoscientific risks. The presence of multiple publics challenges the conceptualization of civic epistemology as homogenously applicable to national/subnational contexts (cf. Jasanoff 2005). Civic epistemology, as a theoretical concept, is helpful in understanding the 'civic' and legally sanctioned engagement of scientific-citizen publics with science but seems to be insufficient to understand the relationship other publics have with science.[30] Democratization of science through the scientific public sphere is therefore crucial but limited, and our analytical lenses should now focus on the alternative sites of engagement opened up by the multiple publics that demand deeper imaginations of democracy beyond the deliberative turn.

Notes

1. *The Hindu* is a national daily published in English. As mentioned elsewhere, the newspaper has a weekly science and technology page that presents what happened in the world of science in a matter-of-fact manner. The people whom I interacted with belonged to the educated middle classes of Kerala who read English dailies like *The Hindu* along with Malayalam newspapers.

2. Ulrich Beck's notion of active subpolitics refers to 'the contestation of taken-for-granted assumptions of modernity by actors outside the system of formal politics' (Holzer and Sørensen 2003: 82).

3. See Ezrahi (1990) for a detailed understanding of the co-evolution of modern science and liberal democratic politics. See also Jasanoff (2004).

4. Italics as in the original.

5. See Rajendran (2001). The journalist who contributed was B. Ajith Babu, sub-editor, *Malalayala Manorama*.

6. See Cloitre and Shinn (1985) and Mellor (2003) for the argument that popular science stabilizes science.

7. The best examples are Anilkumar (2011) and Thomas (2012). See Gregory (2003) and Lewenstein (1995a) for case studies on popular science during scientific controversies.

8. 'We Can't Be Blind to Tamil Nadu's Fears over "DAM 999", Says Supreme Court', *The Hindu*, 25 January 2013, available at http://www.thehindu.com/news/national/we-cant-be-blind-to-tamil-nadus-fears-over-dam-999-says-supreme-court/article4341358.ece (accessed on 11 December 2015).

9. See Pillai (2012) for a detailed analysis of the debates around these two films.

10. The discussion here in no way intends to present a holistic picture of the scientific research system in the region. Only a few significant features revealed by the controversies under study are highlighted. See Sooryamoorthy and Shrum (2004) for a broader analysis of the scientific community in Kerala.

11. The CESS (Thiruvananthapuram), the Centre for Water Resources Development and Management (CWRDM, Kozhikode), the Tropical Botanical Garden and Research Centre (TBGRI, Thiruvananthapuram), the Kerala State Remote Sensing and Environment Centre (KSREC, Thiruvananthapuram), and the RCC (Thiruvananthapuram) are autonomous research centres under the state government. National institutes include the Regional Research Laboratory (RRL), the Central Marine Fisheries Research Institute (CMFRI, Kochi), the Central Plantation Crops Research Institute (CPCRI, Kasaragod), the Meteorological Centre (Thiruvananthapuram), and the Vikram Sarabhai Space Centre (VSSC, Thiruvananthapuram).

12. The Bhabha Atomic Research Centre (BARC, Mumbai), the Central Groundwater Board (CGWB), the Central Soil and Materials Research Station (CSMRS, New Delhi), the Geological Survey of India (GSI, Kolkata), the India Meteorological Department (IMD, New Delhi), and the National Geophysical Research Institute (NGRI, Hyderabad) were the major national research institutes that participated in the controversies under study.

13. The School of Pure and Applied Physics (SPAP) and the School of Applied Life Sciences (SALS).

14. The University of Cochin, established by the Government of Kerala in 1971, changed its name to Cochin University of Science and Technology (CUSAT) in 1986.

15. The institute was established in 1976. Currently it functions as an autonomous institute with university status under the DST, GoI.

16. Personal interviews with the scientists (names withheld) on 27 April 2006.

17. G. Sankar from the CESS was one among them. He continued with his research on the 'piping effect', a geological phenomenon that had been reported from the region after the earthquakes. Dr Louis and Dr Kumar continued with their research on 'alien cells', even after the controversy had ended.

18. The only exception was the collaborative research undertaken by five scientists from three different scientific research institutions during the controversy over unnatural geophysical phenomena. Their findings were published as Singh et al. (2005).

19. As pointed out by the CESS scientists during personal interviews.

20. For detailed studies on regulatory mechanisms in developed countries, see Bijker, Bal, and Hendriks (2009), Hilgartner (2000), and Jasanoff (1990, 2005).

21. Sooryamoorthy and Shrum (2004) in their study of the 'institutional settings of knowledge creation' in Kerala also point to the low research output and institutional weaknesses of the regional research system in Kerala on the basis of surveys conducted between 1994 and 2000. According to them, scientists in Kerala were less cosmopolitan, and there was a recent decline in international collaboration, accompanied with a reduction in intra-organizational relations during the period of study (Sooryamoorthy and Shrum 2004: 213). Their analysis reveals that productivity in terms of scientific publications did not increase between 1994 and 2000 (Sooryamoorthy and Shrum 2004: 217–18).

22. See the concluding remarks of the editors in Irwin and Wynne (1996: 217).

23. Sheila Jasanoff (2005) refers to communitarian, consensus-seeking, and contentious as three types of civic epistemology respectively in the national contexts of Britain, Germany, and the USA respectively. For a comparative view of the civic epistemology of the three countries, see Jasanoff (2005: 259). However, the civic epistemology of Kerala being contentious, manifested through the scientific public sphere, does not imply that it shares all the characteristics with that of the US.

24. See more on this later in the chapter.

25. For an introduction to the Plachimada struggle, see Bijoy (2006) and Raman (2005). For the endosulfan controversy, see Irshad and Joseph (2015) and Varughese (2012).

26. The committee that submitted its report in 2011 is popularly known as the Gadgil Commission after its chairman, Madhav Gadgil. See Kunhikkannan (2013) and Mohan (2014) for more details.

27. The controversy regarding the dumping of urban municipal solid waste at Vilappilsala, a village on the outskirts of the city of Thiruvananthapuram, is a case in point. The villagers are at loggerheads with the city dwellers in their risk perception and views on waste management (Subair 2014). The conflict between these two publics with regards to their differing perceptions of the

risks of waste (and hence conflicts on waste management) is at the heart of the controversy and its deliberation in the scientific public sphere.

28. The 1990s have been marked as the period of emergence of a new dalit and feminist politics in Kerala's public sphere. For more details, see Dasan et al. (2012), Devika and Sukumar (2006), Raj (2013), and Tharu and Satyanarayana (2011).

29. 'Letter to the Editor', *Mathrubhumi*, 18 January 1968, p. 4, quoted in Devika (2008: 174–5).

30. A third category of public along with scientific-citizen publics and quasi-publics could be the 'non-publics' who are at the threshold of the political community and, hence, cannot be included as subjects of politics. The disfigured and genetically mutated bodies of non-publics are with the potential to challenge the very liberal democratic notion of public deliberation of science. For a detailed discussion, see Varughese (2012). What implications do multiple publics and their diverse relationship with science have for regional modernities is a pertinent question. Attending to this problem, however, is beyond the scope of this book. For a theoretical exploration of the idea of regional modernity beyond a reductionist understanding of region as sub-national, see Bose and Varughese (2015).

Appendix

The unusual geophysical incidents reported from Kerala[1]

1. Tremors
2. Cracks on walls
3. Waves in wells
4. Water springing out from tube wells/wells
5. Vanishing wells; well collapses; wells getting filled with mud
6. Draining of water in wells; wells drying up
7. Abnormal rise and fall in water level in wells
8. Bitter/sweet taste of water from wells
9. Colour change in water inside collapsed wells (blue and red)
10. Water springing out from the ground
11. Water bubbling and boiling in wells and paddy fields
12. Houses collapsing into the ground
13. Fumes from wells and ground
14. Fuming hill in Idinjimala, Idukki district
15. Rivulets disappearing into the earth
16. Exploding rocks
17. Tunnel formation/well formation
18. Sound of explosion/humming under the ground
19. Smell of gunpowder associated with other phenomena
20. Fireball appearing in the sky
21. Coloured rain (red, black, yellow, green, and blue)
22. 'Insect-rain' and 'shrimp-rain' (presence of insects and shrimps in rain water)
23. Insects in wells
24. Burning gas coming out of tube wells

25. Trees shedding leaves and drying up even in rainy season
26. Foul smell in sea wind
27. Landslide
28. Sea becomes furious; high tide

Note

1. The geophysical phenomena are referred to here using literal translations of the terms developed by the regional newspapers rather than the scientific terminology employed in technical papers.

References

Abraham, Itty. 1998. *The Making of the Indian Atomic Bomb: Science, Secrecy and the Postcolonial State.* Hyderabad: Orient Longman.

——. 2006. 'The Contradictory Spaces of Postcolonial Techno-Science'. *Economic and Political Weekly* 41(3), 21 January: 210–17.

——. 2012. 'Geopolitics and Biopolitics in India's Natural Background Radiation Zone'. *Science, Technology & Society* 17(1): 105–22.

Achyuthan, A. 2012 [2009]. *Mullaperiyar Imbroglio: Inter-State River Problems.* Second Edition. Thrissur: Kerala Sasthra Sahithya Parishad.

Addlakha, Renu. 2001. 'State Legitimacy and Social Suffering in a Modern Epidemic: A Case Study of Dengue Hemorrhagic Fever in Delhi'. *Contributions to Indian Sociology* (n.s.) 35(2): 151–79.

Adiyodi, Dr. K.G. 2006. *Anweshanam* (Enquiry, Mal.). Thrissur: Kerala Sahitya Akademi.

Anilkumar, A.V. (ed.). 2011. *Mullaperiyar* (Mal.). Kottayam: Papyrus Books.

Arunima, G. 2006. 'Imagining Communities—Differently: Print, Language and the "Public Sphere" in Colonial Kerala'. *The Indian Economic and Social History Review* 43(1): 63–76.

Bagla, Pallava. 2001. 'India Acts on Flawed Cancer Drug Trials'. *Science* 28(September): 2371, 2373.

Bagla, Pallava and Eliot Marshall. 2001. 'Hopkins Reviews Investment in Indian Cancer Drug Trial'. *Science* 293(5532), 10 August: 1024.

Balakrishnan, Kavumbayi. 2007. *Malayala Sasthrasahithya Prasthanam: Oru Padanam* (Popular Science Literary Movement in Malayalam: A Study, Mal.). Thrissur: Kerala Sasthra Sahithya Parishad.

Bandyopadhyay, Jayanta and Vandana Shiva. 1988. 'Political Economy of Ecology Movements'. *Economic and Political Weekly*, 23 (24), 11 June: 1223–32.

Bast, Felix Satej Bhushan, Aijaz Ahmad John, Jackson Achankunju, Nadaraja Panikkar MV, Christina Hametner, and Elfriede Stocker-Wörgötter

2015. 'European Species of Subaerial Green *Alga Trentepohlia annulata* (*Trentepohliales, Ulvophyceae*) Caused Blood Rain in Kerala, India', *Phylogenetics & Evolutionary Biology* 3(1): 144.

Bay, Ancy. 2015. 'At the End of the Story: Popular Fiction, Readership and Modernity in Literary Malayalam'. In *Kerala Modernity: Ideas, Spaces and Practices in Transition*, edited by Satheese Chandra Bose and Shiju Sam Varughese, 92–108. Hyderabad: Orient Blackswan.

Beck, Ulrich. 1992. *Risk Society: Towards a New Modernity*. London, New Bury Park, and New Delhi: Sage Publications.

———. 1997. 'Subpolitics: Ecology and the Disintegration of Institutional Power'. *Organisation & Environment* 10(1), March: 52–65.

———. 2009. *World at Risk*, translated by Ciaran Cronin. Cambridge and Malden, MA: Polity Press.

Bernal, J.D. 1939. *The Social Function of Science*. London: George Routledge & Sons Ltd.

Bhaduri, Saradindu and Aviram Sharma. 2014. 'Public Understanding of Participation in Regulatory Decision-Making: The Case of Bottled Water Quality Standards in India'. *Public Understanding of Science* 23(4): 472–88.

Bharadwaj, Aditya. 2000. 'How Some Indian Baby Makers Are Made: Media Narratives and Assisted Conception in India'. *Anthropology & Medicine* 7(1): 63–78.

Bhatia, Aditi. 2006. 'Critical Discourse Analysis of Political Press Conferences'. *Discourse and Society* 17(2), March: 173–203.

Bijker, Wiebe E., Roland Bal, and Ruud Hendriks. 2009. *The Paradox of Scientific Authority: The Role of Scientific Advice in Democracies*. Cambridge and London: The MIT Press.

Bijoy, C.R. 2006. 'Kerala's Plachimada Struggle: A Narrative on Water and Governance Rights'. *Economic and Political Weekly* 14, October: 4332–9.

Bloor, David. 1991[1976]. *Knowledge and Social Imagery*. Second Edition. Chicago and London: The University of Chicago Press.

Bocking, Stephen. 2010. 'Mobile Knowledge and the Media: The Movement of Scientific Information in the Context of Environmental Controversy'. *Public Understanding of Science* 21(6): 705–23.

Bose, Staheese Chandra and Shiju Sam Varughese. 2015. 'Situating an Unbound Region: Reflections on Kerala Modernity'. In *Kerala Modernity: Ideas, Spaces and Practices in Transition*, edited by Satheese Chandra Bose and Shiju Sam Varughese, 1–24. Hyderabad: Orient Blackswan.

Bourdieu, Pierre. 1975. 'The Specificity of the Scientific Field and the Social Conditions of the Progress of Reason'. *Social Science Information* 14(6): 19–47.

Broman, Thomas. 1998. 'The Habermasian Public Sphere and "Science *in* the Enlightenment"'. *History of Science* 36: 123–49.

Bucchi, Massimiano. 1996. 'When Scientists Turn to the Public: Alternative Routes in Science Communication'. *Public Understanding of Science* 5(4): 375–94.

———. 1998. *Science and the Media: Alternative Routes in Scientific Communication*. London and New York: Routledge.

Calhoun, Craig. 1992. 'Introduction: Habermas and the Public Sphere'. In *Habermas and the Public Sphere*, edited by Craig Calhoun, 1–48. Cambridge and London: The MIT Press.

Callon, Michel. 1986. 'Some Elements of a Sociology of Translation: Domestication of the Scallops and the Fishermen of St Brieuc Bay'. In *Power, Action and Belief: A New Sociology of Knowledge?* edited by John Law, 196–223. London and Boston: Routledge & Kegan Paul.

Calsamiglia, Helena and Teun A. van Dijk. 2004. 'Popularisation Discourse and Knowledge about the Genome'. *Discourse and Society* 15(4), July: 369–89.

Carvalho, Anabela. 2007. 'Ideological Cultures and Media Discourses on Scientific Knowledge: Re-reading News on Climate Change'. *Public Understanding of Science* 16(2): 223–43.

Catone-Huber, Adrienne and Jennifer Smith. N.d. '1967 Earthquake at Koyna Dam, India: The Study of a Reservoir-Induced Earthquake'. Available at http://www.hubcat.org/Adrienne/Koyna/Koyna_home.htm (accessed on 19 January 2007).

Centre for Development Studies (CDS). 1975. *Poverty, Unemployment, and Development Policy: A Case Study of Selected Issues with reference to Kerala*. New York: Department of Economic and Social Affairs, United Nations.

Chakraborty, Amit (ed.). 2001. *Communicating Science: A Book on Popular Presentation of Science*. Calcutta: Miranda.

Christy K.J., Carmel. 2015. 'The Politics of Sexuality and Caste: Looking through Kerala's Public Space'. In *Kerala Modernity: Ideas, Spaces and Practices in Transition*, edited by Satheese Chandra Bose and Shiju Sam Varughese, 126–45. Hyderabad: Orient Blackswan.

Cloitre, M. and Shinn, T. 1985. 'Expository Science: Social, Cognitive and Epistemological Linkages'. In *Expository Science: Forms and Functions of Popularisation*, edited by Terry Shinn and Richard Whitley, 31–60. Dordrecht, Boston, and Lancaster: D. Reidel Publishing Company.

Collins, H.M. (ed.). 1981. 'Knowledge and Controversy: Studies of Modern Natural Science'. *Social Studies of Science*, special issue, 11(1): 3–158.

Collins, H.M. 1987. 'Certainty and the Public Understanding of Science: Science on Television 1'. *Social Studies of Science* 17(4): 689–713.

Cook, Guy, Peter T. Robbins, and Elisa Pieri. 2006. '"Words of Mass Destruction": British Newspaper Coverage of the Genetically Modified Food Debate, Expert and Non-expert Reactions'. *Public Understanding of Science* 15(1): 5–29.

Cottle, Simon. 1998. 'Ulrich Beck, "Risk Society" and the Media: A Catastrophic View?.' *European Journal of Communication* 13(1): 5–32.

Crossley, Nick. 2004. 'On Systematically Distorted Communication: Bourdieu and the Socio-analysis of Publics'. In *After Habermas: New Perspectives on the Public Sphere*, edited by Nick Crossley and John Roberts, 88–112. Oxford and Malden, MA: Blackwell Publishing.

D'Monte, Darryl. 1985. *Temples or Tombs? Industry versus Environment: Three Controversies*. New Delhi: Centre for Science and Environment.

Dahlberg, Lincoln. 2005. 'The Habermasian Public Sphere: Taking Difference Seriously?.' *Theory and Society* 34(2), April: 111–36.

Damodaran, Pradeep. 2014. *The Mullaperiyar Water War: The Dam That Divided Two States*. New Delhi: Rupa Publications India.

Das, Veena. 1995. *Critical Events: An Anthropological Perspective on Contemporary India*. New Delhi: Oxford University Press.

Dasan, M., V. Prathibha, Pradeepan Pampirikkunnu, and C.S. Chandrika (eds). 2012. *The Oxford Anthology of Malayalam Dalit Writing*. New Delhi: Oxford University Press.

Dash, Biswanath. 2014a. 'Public Understanding of Cyclone Warning in India: Can Wind Be Predicted?.' *Public Understanding of Science* 24(8): 1–18.

———. 2014b. 'Science, State and the Public'. *Seminar* 654, February: 27–31.

Davenport, Sally and Shirley Leitch. 2005. 'Agoras, Ancient and Modern, and a Framework for Science-Society Debate'. *Science and Public Policy* 32(2), April: 137–53.

Devika, J. 2007. *En-gendering Individuals: The Language of Re-forming in Twentieth Century Keralam*. Hyderabad: Orient Longman.

———. 2008. *Individuals, Householders, Citizens: Malayalis and Family Planning, 1930–1970*. New Delhi: Zubaan.

Devika, J. and Mini Sukumar. 2006. 'Making Space for Feminist Social Critique in Contemporary Kerala'. *Economic and Political Weekly* 21 October: 4469–75.

Dijk, Teun A. Van. 1988. *News as Discourse*. New Jersey: Lawrence Erlbaum Associates, Inc.

Dryzek, John S. 2000. *Deliberative Democracy and Beyond: Liberals, Critics, Contestations*. Oxford and New York: Oxford University Press.

Durant, Darrin. 2011. 'Models of Democracy in Social Studies of Science'. *Social Studies of Science* 41(5): 691–714.

Dutt, Bharvi and K.C. Garg. 2000. 'An Overview of Science and Technology Coverage in Indian-Language Dailies'. *Public Understanding of Science* 9(2): 123–40.

Elzinga, Aant. 1988. 'Bernalism, Comintern and the Science of Science: Critical Science Movements Then and Now'. In *From Research Policy to Social*

Intelligence: Essays for Stevan Dedijer, edited by Jan Annerstedt and Andrew Jamison, 87–113. Houndmills, Basingstoke, Hampshire, and London: Macmillan Press.

Elzinga, Aant and Andrew Jamison. 1986. 'The Other Side of the Coin: The Cultural Critique of Technology in India and Japan'. In *Technological Developments in China, India and Japan*, edited by E. Baark and Andrew Jamison, 205–51. London: Maxmillan.

Engelhardt, H. Tristram and Arthur L. Caplan. 1987. 'Patterns of Controversy and Closure: The Interplay of Knowledge, Values, and Political Forces'. In *Scientific Controversies: Case Studies in the Resolution and Closure of Disputes in Science and Technology*, edited by H. Tristram Engelhardt and Arthur L. Caplan, 1–23. Cambridge: Cambridge University Press.

Epstein, Steven. 1995. 'The Construction of Lay Expertise: AIDS Activism and the Forging of Credibility in the Reform of Clinical Trials'. *Science, Technology & Human Values* 20(4): 408–37.

———. 1996. *Impure Science: AIDS, Activism, and the Politics of Knowledge*. Berkeley, Los Angeles, and London: University of California Press.

Ezrahi, Yaron. 1990. *The Descent of the Icarus: Science and the Transformation of Contemporary Democracy*. Cambridge: Harvard University Press.

———. 1995. 'Technology and the Illusion of Escape from Politics'. In *Technology, Pessimism and Postmodernism*, edited by Yaron Ezrahi, Everett Mendelsohn, and Howard Segal, 29–37. Amherst: University of Massachusetts Press.

Felt, Ulrike and Maximilian Fochler. 2010. 'Machineries for Making Publics: Inscribing and De-scribing Publics in Public Engagement'. *Minerva* 48(3): 219–39.

Fischer, Frank. 2005. *Citizens, Experts, and the Environment*. Durham: Duke University Press.

Franzen, Martina, Peter Weingart, and Simone Rödder. 2012. 'Exploring the Impact of Science Communication on Scientific Knowledge Production: An Introduction'. In *The Sciences' Media Connection: Public Communication and Its Repercussions*, edited by Simone Rödder, Martina Franzen, and Peter Weingart, 3–14. Dordrecht, Heidelberg, London and New York: Springer.

Fraser, Nancy. 1995. 'Politics, Culture, and the Public Sphere: Toward a Postmodern Conception'. In *Social Postmodernism: Beyond Identity Politics*, edited by Linda Nicholson and Steven Seidman, 287–312. Cambridge: Cambridge University Press.

———. 2003. 'Rethinking the Public Sphere: A Contribution to the Critique of Actually Existing Democracy'. In *Civil Society and Democracy: A Reader*, edited by Carolyn M. Elliott, 83–105. New Delhi: Oxford University Press.

Funtowicz, Silvio O. and Jerome R. Ravetz. 1992. 'Science for the Post-normal Age'. *Futures*, September: 739–55.

Galison, Peter and David J. Stump (eds). 1996. *The Disunity of Science: Boundaries, Contexts, and Power*. Stanford: Stanford University Press.

Gangappa, Rajkumar, Chandra Wickramasinghe, Milton Wainwright, A. Santhosh Kumar, and Godfrey Louis. 2010. 'Growth and Replication of Red Rain Cells at 121°C and Their Red Fluorescence', Proceedings of SPIE Conference 7819: Instruments, Methods, and Missions for Astrobiology XIII, 3–5 August, San Diego, California, USA.

Gangappa, Rajkumar and Stuart I. Hogg. 2013. 'DNA Unmasked in the Red Rain Cells of Kerala', *Microbiology* 159: 107–11.

George, M.V. 2011. *Mullaperiyar Dam: Keralam Neridunna Durantha Bheeshani* (Mullaperiyar dam: The Risk of Disaster Faced by Kerala, Mal.). Kottayam: DC Books.

Gibbons, Michael, Camille Limoges, Helga Nowotny, Simon Schwartzman, Peter Scott, and Martin Trow. 1994. *The New Production of Knowledge: The Dynamics of Science and Research in Contemporary Societies*. London: Sage Publications.

Gieryn, Thomas F. 1995. 'Boundaries of Science'. In *Handbook of Science and Technology Studies*, edited by Sheila Jasanoff, Gerald E. Markle, James C. Peterson, and Trevor Pinch, 393–443. Thousand Oaks, London and New Delhi: Sage Publications.

Goffman, Erving. 1956. *The Presentation of Self in Everyday Life*. Edinburgh: University of Edinburgh Social Science Research Centre.

Govi, K.M. 1998. *Adimudranam: Bharathathilum Malayalathilum* (History of Printing in India with special reference to Malayalam, Mal.). Thrissur: Kerala Sahitya Akademi.

Gregory, Jane. 2003. 'The Popularisation and Excommunication of Fred Hoyle's "Life-from-Space Theory"'. *Public Understanding of Science* 12(1): 25–46.

Guha, Ramachandra. 1988. 'Ideological Trends in Indian Environmentalism'. *Economic and Political Weekly* 23(49): 2578–81.

———. 2013. *The Unquiet Woods: Ecological Change and Peasant Resistance in the Himalaya*, 20th anniversary edition. Ranikhet: Permanent Black.

Guston, David H. 1992. 'The Demise of the Social Contract for Science: Misconduct in Science and Non-modern World', Working Paper No. 19, Programme in Science, Technology and Society, Massachusetts Institute of Technology.

Habermas, Jurgen. 1989. *The Structural Transformation of the Public Sphere: An Inquiry into a Category of Bourgeois Society*, translated by Thomas Burger. UK: Polity Press.

———. 1992. 'Further Reflections on the Public Sphere'. In *Habermas and the Public Sphere*, edited by Craig Calhoun, 421–61. Cambridge and London: The MIT Press.

———. 2000. 'The Public Sphere'. In *Readings in Contemporary Political Sociology*, edited by Kate Nash, 288–94. Massachusetts and Oxford: Blackwell Publishing.

Hagendijk, R.P. 2004. 'The Public Understanding of Science and Public Participation in Regulated Worlds'. *Minerva* 42(1): 41–59.

Heidegger, Martin. 1977. *The Question Concerning Technology and Other Essays*. New York and London: Garland Publishing, Inc.

Hepp, Andreas. 2012. 'Mediatisation and the "Moulding Force" of the Media'. *Communications* 37(1): 1–28.

Hess, David J. 2011. 'To Tell the Truth: On Scientific Counter Publics'. *Public Understanding of Science* 20(5): 627–41.

Hilgartner, Stephen. 1990. 'The Dominant View of Popularisation: Conceptual Problems, Political Uses'. *Social Studies of Science* 20(3): 519–39.

———. 2000. *Science on Stage: Expert Advice as Public Drama*. Stanford: Stanford University Press.

Holzer, Boris and Mads P. Sørensen. 2003. 'Rethinking Subpolitics: Beyond the "Iron Cage" of Modern Politics?' *Theory, Culture & Society* 20(2): 79–102.

Irshad, S. Mohammed and Jacquleen Joseph. 2015. 'An Invisible Disaster: Endosulfan Tragedy of Kerala'. *Economic and Political Weekly* 50(11), 14 March: 61–5.

Irwin, Alan. 1995. *Citizen Science: A Study of People, Expertise and Sustainable Development*. London and New York: Routledge.

———. 2001. 'Constructing the Scientific Citizen: Science and Democracy in the Biosciences'. *Public Understanding of Science* 10(1): 1–18.

Irwin, Alan and Brian Wynne (eds). 1996. *Misunderstanding Science? The Public Reconstruction of Science and Technology*. Cambridge: Cambridge University Press.

Irwin, Alan and Mike Michael. 2003. *Science, Social Theory and Public Knowledge*. Maidenhead and Philadelphia: Open University Press.

Isaac, T.M. Thomas and B. Ekbal. 1988. *Science for Social Revolution: The Experience of Kerala Sastra Sahitya Parishat*. Trichur: Kerala Sastra Sahitya Parishat.

Isaac, T.M. Thomas, Richard W. Franke, and Pyaralal Raghavan. 1998. *Democracy at Work in an Industrial Cooperative: The Story of Kerala Dinesh Beedi*. Ithaca and London: ILR Press, an Imprint of Cornell University Press.

Jacob, John C. 2009. *Harithadarshanam: Johnsiyude Athmakatha* (Green Vision: Autobiography of John C., Mal.). Kozhikode: Mathrubhumi Books.

Jacob, Shaji. 2009. *Janapriya Samskaram: Charithravum Sidhanthavum* (Popular Culture: History and Theory, Mal.). Kozhikkode: Mathrubhumi Books.

James, Saji. 2010. *Silent Valley: Oru Paristhithi Samarathinte Charithram* (Silent Valley: History of an Ecological Struggle, Mal.). Kottayam: DC Books.

Jasanoff, Sheila. 1987. 'Contested Boundaries in Policy-Relevant Science'. *Social Studies of Science* 17(2):195–230.

———. 1990. *The Fifth Branch: Science Advisers as Policy Makers*. Cambridge and London: Harvard University Press.

———. 2003a. '(No?) Accounting for Expertise'. *Science and Public Policy* 30(3): 157–62.

———. 2003b. 'Breaking the Waves in Science Studies: Comment on H.M. Collins and Robert Evans, "The Third Wave of Science Studies"'. *Social Studies of Science* 33(3): 389–400.

———. 2003c. 'Technologies of Humility: Citizen Participation in Governing Science'. *Minerva* 41(3): 223–44.

——— (ed.). 2004. *States of Knowledge: The Co-production of Science and Social Order.* London and New York: Routledge.

———. 2005. *Designs on Nature: Science and Democracy in Europe and the United States.* Princeton and Oxford: Princeton University Press.

Jasanoff, Sheila, Gerald E. Markle, James C. Peterson, and Trevor Pinch (eds). 1995. *Handbook of Science and Technology Studies.* London: Sage Publications.

Jeffrey, Robin. 1992. *Politics, Women and Well-Being: How Kerala Became 'A Model'.* New Delhi: Oxford University Press.

———. 1997a. 'Advertising and Indian-Language Newspapers: How Capitalism Supports (Certain) Cultures and (Some) States, 1947–'96'. *Pacific Affairs* 70(1): 57–84. Also in Robin Jeffrey, *Media and Modernity: Communications, Women, and the State in India* (Ranikhet: Permanent Black, 2010), 166–99.

———. 1997b. 'Indian Language Newspapers 1: Malayalam: "The Day-to-Day Social Life of the People..."' *Economic and Political Weekly* 32(1–2), 4–11 January: 18–21.

———. 2000. *India's Newspaper Revolution: Capitalism, Politics and the Indian Language Press.* New Delhi: Oxford University Press.

———. 2003. 'Breaking News'. *The Little Magazine* 4(2): 6–10.

———. 2010. *Media and Modernity: Communications, Women, and the State in India.* Ranikhet: Permanent Black.

Johnson, James H. 1992. 'Musical Experience and the Formation of a French Musical Public'. *Journal of Modern History* 64, June: 191–226.

Kellner, Douglas. 2000. 'Habermas, the Public Sphere, and Democracy: A Critical Intervention'. Available at http://www.gseis.ucla.edu/faculty/kellner/kellner.html (accessed on 14 October 2004).

Kerala Sasthra Sahithya Parishad (KSSP). 1984. *Science as Social Activism: Reports and Papers on the People's Science Movements in India.* Trivandrum.

———. 1993. *Parishad Pinnitta Muppathu Varshangal* (The Thirty Years of the KSSP, Mal.). Kozhikode.

Kerr, Anne, Sarah Cunningham-Burley, and Richard Tutton. 2007. 'Shifting Subject Positions: Experts and Lay People in Public Dialogue'. *Social Studies of Science* 37(3): 385–411.

Knorr-Cetina, Karrin D. 1981. *The Manufacture of Knowledge: An Essay on the Constructionist and Contextual Nature of Science.* Oxford: Pergamon Press.

Koopmans, Ruud. 2004. 'Movements and Media: Selection Processes and Evolutionary Dynamics in the Public Sphere'. *Theory and Society* 33(3–4), June–August: 367–91.

Krishnakumar, R. 2001. 'Ethics on Trial'. *Frontline* 18(16), 4–17 August: 123–26.

———. 2005. 'Erimos Has Broadened the Research: Interview with Ru Chih C. Huang, Biology Professor of Johns Hopkins University'. *Frontline* 22(25). Available at http://www.frontline.in/static/html/fl2225/stories/20051216004602400.htm (accessed on 02 November 2014).

Kuhn, Thomas. 1996[1962]. *The Structure of Scientific Revolutions*. Chicago and London: The University of Chicago Press.

Kumar, Santhosh, Chandra Wickramasinghe, and Godfrey Louis. 2013. 'A Comparative Study of Trentepohlia and Red Cells'. *International Journal of Recent Scientific Research* 4(8): 1205–9.

Kumar, Udaya. 2007. 'The Public, the State and New Domains of Writing: On Ramakrishna Pillai's Conception of Literary and Political Expression'. *Tapasam*, January and April: 413–41.

———. 2013. 'Abhiruchiyum Achadakkavum: Anthappayiye Innu Vayikkumpol' (Taste and Discipline: Reading Anthappayi Today). In *Naluperiloruthan adhava Nadakadyam Kavithvam* (Almost Everyone, or Poets Become Playwrights, Mal.), edited by C. Anthappayi, 11–40. Thiruvananthapuram: Chintha Publishers.

Kumar, V. Sasi, S. Sampath, C.N. Mohanan, and T.K. Abraham. 2002. 'Coloured Rain Falls in Kerala, India'. *Eos* 83(31): 335.

Kunhikkannan, T.P. 2013. *Gadgil Reportum Keralavikasanavum* (Gadgil Report and Kerala Development, Mal.). Kozhikode: Mathrubhumi Books.

———. 2014. *Sasthram Samarayudhamakumpol* (Science as a Weapon for Struggle, Mal.). Thrissur: Kerala Sasthra Sahithya Parishad.

Kunhikrishnan, K. 2006. 'Fast Paced Media Scenario'. *Kerala Calling*, November: 31–2. Also available at http://www.kerala.gov.in/kercalnovmbr06/pg31-32.pdf (accessed on 15 January 2008).

Latour, Bruno. 1987. *Science in Action: How to Follow Scientists and Engineers through Society*. Cambridge and Massachusetts: Harvard University Press.

———. 1988. *The Pasteurisation of France*, translated by Alan Sheridan and John Law. Cambridge, Massachusetts, and London: Harvard University Press.

———. 1996. *Aramis or the Love of Technology*, translated by Catherine Porter. Cambridge, Massachusetts, and London: Harvard University Press.

Latour, Bruno and Steve Woolgar. 1979. *Laboratory Life: The Social Construction of Scientific Facts*. Beverly Hills and London: Sage Publications.

Laxman, R.K. 1982. *Science Smiles*. Bombay: India Book House Pvt. Ltd.

Leach, Melissa, Ian Scoones, and Brian Wynne (eds). 2005. *Science and Citizens: Globalisation and the Challenge of Engagement*. London and New York: Zed Books.

Lee, Francis L.F. and Angel M.Y. Lin. 2006. 'Newspaper Editorial Discourse and the Politics of Self-Censorship in Hong Kong'. *Discourse and Society* 17(3): 331–8.

Levina, Marina. 2009. 'Exploring Epistemic Boundaries between Scientific and Popular Cultures'. *Spontaneous Generations: A Journal for the History and Philosophy of Science* 3(1): 105–12.

Lewenstein, Bruce V. 1995a. 'From Fax to Facts: Communication in the Cold Fusion Saga'. *Social Studies of Science* 25(3): 403–36.

———. 1995b. 'Science and the Media'. In *Handbook of Science and Technology Studies*, edited by Sheila Jasanoff et al., 343–60. London: Sage Publications.

Livingstone, Sonia (ed.). 2005. *Audiences and Publics: When Cultural Engagement Matters for the Public Sphere*. Bristol and Portland: Intellect Books.

Livingstone, Sonia. 2009. 'Foreword: Coming to Terms with "Mediatisation"'. In *Mediatisation: Concept, Changes, Consequences*, edited by Knutt Lundby, ix–xi. New York: Peter Lang.

Louis, Godfrey and A. Santhosh Kumar. 2003. 'New Biology of Red Rain Extremophiles Prove Cometary Panspermia'. Available at http//:arxiv.org/abs/astro-ph/0312639 (accessed on 20 October 2006).

———. 2006. 'The Red Rain Phenomenon of Kerala and Its Possible Extra-terrestrial Origin'. *Astrophysics and Space Science*, 302(1–4): 175–87. Available at http//:arxiv.org/abs/astro-ph/0310120 (accessed on 20 October 2006).

Lövbrand, Eva, Roger Pielke, Jr., and Silke Beck. 2011. 'A Democracy Paradox in Studies of Science and Technology'. *Science, Technology and Human Values* 36(4): 474–96.

Luhmann, Niklas. 1995. *Social Systems*, translated by John Bednarz, Jr., with Dirk Baecker. Stanford: Stanford University Press.

———. 2000. *The Reality of the Mass Media*, translated by Kathleen Cross. Stanford: Stanford University Press.

Lukose, Ritty A. 2010. *Liberalisation's Children: Gender, Youth, and Consumer Citizenship in Globalising India*. Hyderabad: Orient Blackswan.

Lundby, Knutt (ed.). 2009. *Mediatisation: Concept, Changes, Consequences*. New York: Peter Lang.

Mangathil, Sasidharan. 2008. *Mullaperiyar Anakkettum Kerlathinte Bhaviyum: Anweshana Rekha* (Mullaperiyar Dam and the Future of Kerala: An Enquiry, Mal.). Kozhikode: Mathrubhumi Books.

Martin, Brian. 1988. 'Analysing the Fluoridation Controversy: Resources and Structures'. *Social Studies of Science* 18(2): 331–63.

Martin, Brian and Evelleen Richards. 1995. 'Scientific Knowledge, Controversy, and Public Decision Making'. In *Handbook of Science and Technology Studies*, edited by Sheila Jasanoff et al., 506–26. London: Sage Publications.

Mazzonetto, Marzia. 2005. 'Science Communication in India: Current Situation, History and Future Developments'. *Journal of Science Communication* 4(1), March: 1–6.

McNair, Brian. 2000. *Journalism and Democracy: An Evaluation of the Political Public Sphere*. London and New York: Routledge.

Mehta, Nalin. 2008. *India on Television: How Satellite News Channels Have Changed the Way We Think and Act*. New Delhi: Harper Collins Publishers India with the India Today Group.

Mellor, Felicity. 2003. 'Between Fact and Fiction: Demarcating Science from Non-Science in Popular Physics Books'. *Social Studies of Science* 33(4): 509–38.

Mendelsohn, Everett. 1987. 'The Political Anatomy of Controversy in the Sciences'. In *Scientific Controversies: Case Studies in the Resolution and Closure of Disputes in Science and Technology*, edited by H. Tristram Engelhardt and Arthur L. Caplan, 93–124. Cambridge: Cambridge University Press.

Menon, Dilip M. 1994. *Caste, Nationalism and Communism in South India: Malabar, 1900–1948*. Cambridge: Cambridge University Press.

Merton, Robert K. 1968. 'The Matthew Effect in Science', *Science*, (n.s.), 159 (3810): 56–63.

Michael, Mike. 1996. 'Ignoring Science: Discourses of Ignorance in the Public Understanding of Science'. In *Misunderstanding Science? The Public Reconstruction of Science and Technology*, edited by Alan Irwin and Brian Wynne, 107–25. Cambridge: Cambridge University Press.

———. 2002. 'Between Science and the Public'. *Science as Culture* 11(1): 115–20.

Miller, Steve. 2001. 'Public Understanding of Science at the Crossroads'. *Public Understanding of Science* 10(1): 115–20.

Mirowski, Philip and Robert Van Horn. 2005. 'The Contract Research Organisation and the Commercialisation of Scientific Research'. *Social Studies of Science* 35(4): 503–48.

Mohan, Manila C. (ed.). 2014. *Madhav Gadgilum Paschimakhatta Samrakshanavum* (Madhav Gadgil and the Conservation of Western Ghats, Mal.). Kozhikode: Mathrubhumi Books.

Mohan, P. Sanal. 2015. *Modernity of Slavery: Struggle against Caste Inequality in Colonial Kerala*. New Delhi: Oxford University Press.

Mukherjee, Saumitra. 2008. 'Cosmic Influence on the Sun–Earth Environment', *Sensors* 8(12): 7736–52.

Nair, J. Rajasekharan. 1999. *Spies from Space: The ISRO Frame-Up*. Delhi: Konark Publishers.

Nair, M. Krishnan. 2013. *Njanum RCCyum: Cancerinoppam Nadanna Oru Doctorude Anubhavangal* (RCC and I: Experiences of a Doctor Who Walked with Cancer, Mal.). Kottayam: DC Books.

Nair, Tara S. 2003a. 'Growth and Structural Transformation of Newspaper Industry in India: An Empirical Investigation'. *Economic and Political Weekly* 38(39): 4182–9.

———. 2003b. 'Jeffrey's Reading of Malayalam Press: A Blindfold Slab?' *Economic and Political Weekly* 30(38–9): 4182–5.

Nambiar, Prithi. 2014. *Media Construction of Environment and Sustainability in India*. New Delhi: Sage Publications.

Namboothirippad, M.C. 2014. *Njan Orkkunnu* (I recollect, Mal.). Thrissur: Kerala Sasthra Sahithya Parishad.

Nandy, Ashis. 1988. 'Introduction: Science as a Reason of State'. In *Science, Hegemony and Violence: A Requiem for Modernity*, edited by Ashis Nandy, 1–23. New Delhi: Oxford University Press.

Neidhardt, Friedhelm. 1993. 'The Public as a Communication System'. *Public Understanding of Science* 2(4): 339–50.

Nelkin, Dorothy. 1995a. 'Science Controversies: The Dynamics of Public Disputes in the United States'. In *Handbook of Science and Technology Studies*, edited by Sheila Jasanoff et al., 444–56. London: Sage Publications.

———. 1995b. *Selling Science: How the Press Covers Science and Technology*, revised edition. New York: W.H. Freeman and Company.

Neresini, Frederico. 2000. 'And Man Descended from the Sheep: The Public Debate on Cloning in the Italian Press'. *Public Understanding of Science* 9(4): 359–82.

Ninan, Sevanti. 2007. *Headlines from the Heartland: Reinventing the Hindi Public Sphere*. Los Angeles, London, New Delhi, and Singapore: Sage Publications.

Nori, Miyake, Takafumi Matsui, Jamie Wallis, Daryl Wallis, Anil Saranayake, and Keerithi Wickramarathen. 2013. 'Discovery of Uranium in Outer Coat of Sri Lankan Red Rain Cells'. *Journal of Cosmology* 22(4): 1–8.

Nowotny, H., P. Scott, and M. Gibbons. 2001. *Rethinking Science: Knowledge and the Public in an Age of Uncertainty.* Cambridge: Polity Press.

———. 2003. 'Introduction: "Mode 2" Revisited: The New Production of Knowledge'. *Minerva* 41(3): 179–94.

O'Mahony, P. and Mike S. Schäfer. 2005. 'The "Book of Life" in the Press: Comparing German and Irish Media Discourse on Human Genome Research'. *Social Studies of Science* 35(1): 99–130.

Padmanabhan, B.S. 2004. 'Science for Social Progress'. *Frontline* 21(9), 24 April–7 May. Available at http://www.frontline.in/static/html/fl2109/stories/20040507003409900.htm (accessed on 28 November 2015).

Parameswaran, M.P. 2002. 'Sasthrasahithya Parishathum Kerala Samoohavum' (KSSP and Kerala Society). In *Nammude Sahityam, Nammude Samooham: 1901–2000*, Vol. III (Our Literature, Our Society: 1901–2000, Mal.), edited by M.N. Vijayan, 119–51. Thrissur: Kerala Sahitya Akademi.

————. 2008. *Janakeeya Sasthra Prasthanam* (People's Science Movement, Mal.). Thrissur: Kerala Sasthra Sahithya Parishad.

Patairiya, Manoj K. and Maria I. Nogueira. 2011. *Sharing Science: India–Brazil Dialogue on Public Communication of Science, Technology, Culture and Society*. National Council for Science and Technology Communication. New Delhi, Sao Polo, and Lucknow: National Council for Science and Technology Communication (Department of Science and Technology, Government of India), Institute of Biomedical Sciences (Universidade de São Paulo), and Indian Science Communication Society.

Pavanan. 2002. 'Yukthivadiprasthanavum Keraleeyajeevithavum' (The Rationalist Movement and Social Life in Kerala, Mal.). In *Nammude Sahityam, Nammude Samooham: 1901–2000*, Vol. III (Our Literature, Our Society: 1901–2000, Mal.), edited by M.N. Vijayan, 44–79. Thrissur: Kerala Sahitya Akademi.

Perez-Lugo, Marla. 2004. 'Media Uses in Disaster Situations: A New Focus on the Impact Phase'. *Sociological Inquiry* 74(2), May: 210–25.

Peters, Hans Peter, Dominique Brossard, Suzanne de Cheveigne, Sharon Dunswoody, Monika Kallfass, Steve Miller, and Shoji Tsuchida. 2008. 'Science–Media Interface: It's Time to Reconsider'. *Science Communication* 30(2), December: 266–76.

Petryna, Adriana. 2002. *Life Exposed: Biological Citizens after Chernobyl*. Princeton: Princeton University Press.

Pillai, Meena T. 2012. 'The Mullaperiyar Dam: Risking Media, Mediating Risk'. *Journal of Creative Communications* 7(1–2): 31–52.

Pillai, P. Govinda. 2003. 'Library Movement and Development Process in Kerala'. In *Information Technology for Participatory Development*, edited by Raman R. Nair, 104–10. New Delhi: Concept Publishers.

Plesner, Ursula. 2010. 'The Performativity of "Media Logic" in the Mass Mediation of Science'. *Public Understanding of Science* 21(6): 674–88.

Pounds, Gabrina. 2006. 'Democratic Participation and Letters to the Editor in Britain and Italy'. *Discourse and Society* 17(1): 29–63.

Prasad, Amit. 2009. 'Capitalising Disease: Biopolitics of Drug Trials in India'. *Theory, Culture & Society* 26(5): 1–29.

Prasad, M.K. and Shabil Krishnan. 2016. 'Silent Valliye Manassilakkan Parishad Patthuvarshameduthu' (Parishad Has Taken Ten Odd Years to Understand Silent Valley, Mal.). *Mathrubhumi Weekly* 93(45), 24–30 January: 30–41.

Prasad, M.K., M.P. Parameswaran, V.K. Nair, K.N. Syamasundaran, and K.P. Kannan. 1979. *The Silent Valley Hydroelectric Project: A Techno-economic and Socio-political Assessment*. Report published by the Health and Environment Brigade of KSSP. Trivandrum: Kerala Sasthra Sahithya Parishad.

Radhakrishnan, R. and Joji Koottummel. 2010. *Silent Valley: Cheruthunilppinte Naalvazhi* (Silent Valley: Chronicles of Resistance, Mal.). Thrissur: Kerala Sasthra Sahithya Parishad.

Radhakrishnan, Ratheesh. 2005. 'P.E. Usha, Hegemonic Masculinities and the Public Domain in Kerala: On the Historical Legacies of the Contemporary'. *Inter-Asia Cultural Studies* 6(2): 187–208.

Raghavan, Puthuppalli. 1985. *Kerala Pathrapravarthanacharithram* (History of Journalism in Kerala, Mal.). Thrissur: Kerala Sahitya Akademi.

Raina, Dhruv. 1993. 'Public Understanding of Science: Why Scientism Beats a Retreat'. *Economic and Political Weekly*, October 16: 2258–61.

———. 1997. 'Evolving Perspectives on Science and History: A Chronicle of Modern India's Scientific Enchantment and Disenchantment (1850–1980)'. *Social Epistemology* 11(1): 3–24.

———. 1999. 'Science and Its Publics'. *India International Quarterly* 26(2), Summer: 42–53.

———. 2007. 'Science since Independence'. In *India 60: Towards a New Paradigm*, edited by Ira Pande, 182–94. New Delhi: Harper Collins Publishers India.

Raina, Dhruv and Ashok Jain. 1997. 'Big Science and the University in India'. In *Science in the Twentieth* Century, edited by John Krige and Dominique Pestre, 859–77. Australia: Harwood Academic Publishers.

Raina, Dhruv and S. Irfan Habib. 2004. *Domesticating Modern Science: A Social History of Science and Culture in Colonial India.* New Delhi: Tulika Books.

Raj, Rekha. 2013. 'Dalit Women as Political Agents: A Kerala Experience'. *Economic and Political Weekly* 48(18), 4 May: 56–63.

Rajagopal, Arvind (ed.). 2009. *The Indian Public Sphere: Readings in Media History.* New Delhi: Oxford University Press.

Rajagopal, Arvind. 2001. *Politics after Television.* Cambridge: Cambridge University Press.

Rajendran, C.P. (guest ed.). 2001. *Bhookampam* (Earthquake, Mal.). Thiruvananthapuram: Kerala Bhasha Institute.

Ramakrishnan, E.V. 2000. 'Varthamanappathrangaludeyum Achadiyanthrangaludeyum Vyapanathode Malayaliyude Sahithyasankalpangalilum Bhashavyavaharangalilum Sambhavicha Mattangal' (Transformations in Literary Imagination and Linguistic Discourses in Malayalam since the Spread of Newspapers and Printing Presses, Mal.). In *Nammude Sahityam, Nammude Samooham: 1901–2000* (Our Literature, Our Society: 1901–2000, Mal.), edited by M.N. Vijayan, 480–504. Thrissur: Kerala Sahitya Akademi.

Raman, K. Ravi. 2005. 'Corporate Violence, Legal Nuances and Political Ecology: Cola War in Plachimada'. *Economic and Political Weekly*, 18 June: 2481–6.

Rastogi, B.K. 2001. 'Erattupetta Earthquake of 12 December 2000 and Seismicity of Kerala', *Journal of Geological Society of India*, 57(3): 273–4.

Raza, Gauhar, Bharvi Dutt, and Surjit Singh. 1991. *Scientific Attitude among People: Report of a Survey Conducted at Allahabad during Kumbh Mela in 1989*, Vols I and II. New Delhi: NISTADS.

Raza, Gauhar, Bharvi Dutt, and Surjit Singh. 1996. *Plague, Media and People*. New Delhi: NISTADS.

———. 1997. 'Kaleidoscoping Public Understanding of Science on Hygiene, Health and Plague: A Survey in the Aftermath of a Plague Epidemic in India'. *Public Understanding of Science* 6(3): 247–67.

Raza, Gauhar, Bharvi Dutt, Surjit Singh, and A. Wahid. N.d. 'Prototype of the Forms of Scientific Cognition: A Survey of Cultural Attitudes to Natural Phenomena'. A Report based on the survey conducted at Mangolpuri, a resettlement colony in Delhi, 1991. New Delhi: NISTADS.

Raza, Gauhar, Surjit Singh, and Bharvi Dutt. 2000. 'Public Understanding of Science in Complex Cultural Structures'. *Journal of Scientific & Industrial Research* 59(2): 460–70.

———. 2002. 'Public, Science and Cultural Distance'. *Science Communication* 23(3): 293–309.

———. 2004. 'Cultural Distance between People's Worldview and Scientific Knowledge in the Area of Public Health'. *Journal of Science Communication* 3(4): 1–8.

Raza, Gauhar, Surjit Singh, and Rajesh Shukla. 2009. 'Relative Cultural Distance and Public Understanding of Science'. *Science, Technology & Society* 14(2): 269–87.

Raza, Gauhar, Surjit Singh, Bharvi Dutt, and Jagdish Chander. 1996. *Confluence of Science and People's Knowledge at the* Sangam, *Allahabad*. New Delhi: NISTADS.

Richardson, J.E. and B. Franklin. 2003. '"Dear Editor": Race, Readers' Letters and the Local Press'. *The Political Quarterly* 74(2), April–June: 184–92.

Rödder, Simone. 2011. 'Science and the Mass Media: "Medialisation" as a New Perspective on an Intricate Relationship'. *Sociology Compass* 5(9): 834–45.

Rödder, Simone and Mike S. Schäfer. 2010. 'Repercussion and Resistance: An Empirical Study on the Interrelation between Science and Mass Media'. *Communications* 35(3): 249–67.

Rödder, Simone, Martina Franzen, and Peter Weingart (eds). 2012. *The Sciences' Media Connection: Public Communication and Its Repercussions*. Dordrecht, Heidelberg, London, and New York: Springer.

Roy, Rahul. 2011. *Dam 999: A Novel*. Bhopal: Indra Publishing House.

Saeed, Saima. 2013. *Screening the Public Sphere: Media and Democracy in India*. London, New York, and New Delhi: Routledge.

Sahay, Uday (ed.). 2006. *Making News: Handbook of the Media in Contemporary India*. New Delhi: Oxford University Press.

Salwi, Dilip M. 2002. *Science in the Indian Media: A Blueprint for the New Millennium*. New Delhi: Vigyan Prasar.

Sam, N. 2003. *Malayala Pathra Pravarthanam Pathonpatham Noottandil* (Journalism in Malayalam Language in the Nineteenth Century, Mal.). Kottayam: DC Books.

Sampath, S., T.K. Abraham, V. Sasi Kumar, and C.N. Mohanan. 2001. 'Coloured Rain: A Report on the Phenomenon'. Thiruvananthapuram: Centre for Earth Science Studies and Tropical Botanic Garden and Research Institute, mimeo. Also available at http://web.archive.org/web/20060613135746/http://www.geocities.com/iamgoddard/Sampath2001.pdf (accessed on 27 March 2016).

Schäfer, Mike S. 2009. 'From Public Understanding to Public Engagement: An Empirical Assessment of Changes in Science Coverage'. *Science Communication* 30(4): 475–505.

———. 2014. 'The Media in the Labs, and the Labs in the Media: What We Know about the Mediatisation of Science'. In *Mediatisation of Communication: Volume 21 of the Handbooks of Communication Science*, edited by Knut Lundby, 571–94. Berlin: de Gruyter Mouton.

Secko, David M., Elyse Amend, and Terrine Friday. 2013. 'Four Models of Science Journalism'. *Journalism Practice* 7(1): 62–80.

Shah, Esha. 2011. '"Science" in the Risk Politics of Bt Brinjal'. *Economic and Political Weekly* 46(31), 30 July: 31–8.

Shapin, Steven. 1974. 'The Audience for Science in Eighteenth Century Edinburgh'. *History of Science* 12(2): 95–121.

———. 1990. 'Science and the Public'. In *Companion to the History of Modern Science*, edited by R.C. Olby et al., 990–1007. London and New York: Routledge.

———. 1994. *A Social History of Truth: Civility and Science in Seventeenth-Century England*. Chicago and London: The University of Chicago Press.

Siddeek, M.A. 2010. *Jalabhramangalil Njan* (I, in the Phantasmic Waters, Mal.). Thrissur: Green Books.

Singh, N.H., J. Mathai, V.N. Neelakandan, D. Shankar, and V.P. Singh. 2005. 'A Database on Occurrence Patterns of Unusual Geological Incidents in Southwest Peninsular India and Its Implication on Future Seismic Activity'. *Acta Geodaetica et Geophysica Hungarica* 40(1): 69–88.

Simon, Bart. 2001. 'Public Science: Media Configuration and Closure in the Cold Fusion Controversy'. *Public Understanding of Science* 10(4): 383–402.

Sismondo, Sergio. 2004. *An Introduction to Science and Technology Studies*. Malden, MA, Oxford, and Victoria: Blackwell Publishing.

Sooryamoorthy, R. and Wesley Shrum. 2004. 'Is Kerala Becoming a Knowledge Society? Evidence from the Scientific Community'. *Sociological Bulletin* 53(2), May–August: 207–21.

Sreedharan, K. (ed.). 1990. *Oorjavivadam* (The Energy Controversy, Mal.). Kozhikode: Kerala Sasthra Sahithya Parishad.

Sreekumar, Sharmila. 2009. *Scripting Lives: Narratives of 'Dominant Women' in Kerala*. Hyderabad: Orient Blackswan.

State Planning Board. 2006. *Economic Review 2005*. Trivandrum: Government of Kerala.

Stichweh, Rudolf. 1992. 'The Sociology of Scientific Disciplines: On the Genesis and Stability of the Disciplinary Structure of Modern Science', *Science in Context* 5(1), March: 3–15.

———. 1996. 'Science in the System of World Society'. *Social Science Information* 35(2): 327–40.

———. 2003. 'The Multiple Publics of Science: Inclusion and Popularisation'. *Sociale Systeme: Zeitschrift fur Soziologische Theorie* 9(2): 210–20.

———. 2015a. 'Analysing Linkages between Science and Politics: Transformations of Functional Differentiation in Contemporary Society'. In *Interfaces of Science and Policy and the Role of Foundations, Conference Report*, edited by Stiftung Mercator, 38–47. Essen.

———. 2015b. 'Transformations in the Interrelation between Science and Nation States: The Theoretical Perspective of Functional Differentiation'. In *Legitimizing Science: National and Global Public, 1800–2010*, edited by Andreas Franzmann, Axel Jansen, and Peter Munte, 35–48. Frankfurt and New York: Campus Verlag.

Subair, K. 2014. 'Public Deliberation over Waste: A Case Study of Vilappilsala Solid Waste Treatment Plant Controversy, Kerala'. Unpublished MPhil dissertation submitted to the Central University of Gujarat, Gandhinagar.

Tharu, Susie and K. Satyanarayana (eds). 2011. *No Alphabet in Sight: New Dalit Writing from South India: Dossier I: Tamil and Malayalam*. New Delhi: Penguin Books.

Thomas, Justice K.T. 2012. *Mullaperiyar Dam: Chila Velippeduthalukal* (Mullaperiyar Dam: Some Disclosures, Mal.). Kottayam: DC Books.

Turner, Stephan. 2001. 'What Is the Problem with Experts?' *Social Studies of Science* 31(1): 123–49.

Unnikrishnan, E. 2008. 'Silent Valley Prakshobham Thudangiyathaaru?' (Who Started the Silent Valley Struggle?, Mal.). *Mathrubhumi Weekly*, 7–13 December, pp. 8–15.

Vadavathoor, Anil Kumar. 2001. *Science Journalism: Vikasavum Parinamavum* (Science Journalism: Growth and Evolution, Mal.). Thiruvananthapuram: Kerala Language Institute.

Varughese, Shiju Sam. 2008. 'Science and Religion Dialogue: Insights from "Public Understanding of Science"'. In *Dancing to Diversity: Science-Religion Dialogue in India*, edited by Kuruvilla Pandikkattu S.J., 13–25. New Delhi: Serials Publishers.

———. 2011. 'Media and Public Controversies over Science: A Case from Kerala, India'. *Spontaneous Generations: A Journal of History and Philosophy of Science* 5(1): 36–43.

———. 2012. 'Where Are the Missing Masses? The Quasi-publics and Non-publics of Technoscience'. *Minerva* 50(2): 239–54.

———. 2014. 'The Public Life of Expertise'. *Seminar* 654, February: 21–6.

———. 2015a. 'Colonial Intellectuals, Public Sphere and the Promises of Modernity: Reading *Parangodeeparinayam*'. In *Kerala Modernity: Ideas, Spaces and Practices in Transition*, edited by Satheese Chandra Bose and Shiju Sam Varughese, 41–58. Hyderabad: Orient Blackswan.

———. 2015b. 'The State-Technoscience Duo in India: A Brief History of a Politico-epistemological Contract'. In *Legitimising Science: National and Global Public, 1800–2010*, edited by Andreas Franzmann, Axel Jansen, and Peter Munte, 137–56. Frankfurt and New York: Campus Verlag.

Venkiteshwaran, C.S. 2014. *Television Padanangal: Nammeyokkeyum Bandhicha Saadhanam…* (Television Studies: The Thing That Binds All of Us.., Mal.). Kozhikode: Mathrubhumi Books.

Vilanilam, J.V. 1993. *Science Communication and Development*. New Delhi, Newbury Park, and London: Sage Publications.

Visvanathan, Shiv. 2007. 'Between Cosmology and System: The Heuristics of a Dissenting Imagination'. In *Another Knowledge is Possible: Beyond Northern Epistemologies*, edited by Boaventura de Sousa Santos, 182–218. London and New York: Verso.

Visvanathan, Shiv and Chandrika Parmar. 2002. 'A Biotechnology Story: Notes from India'. *Economic and Political Weekly* 37(27): 2714–24.

Wang, Jessica. 1999. 'Merton's Shadow: Perspectives on Science and Democracy since 1940'. *Historical Studies in the Physical and Biological Sciences* 30(1): 279–306.

Weingart, Peter. 1998. 'Science and the Media'. *Research Policy* 27(8), December: 869–79.

———. 1999. 'Scientific Expertise and Political Accountability: Paradoxes of Science in Politics'. *Science and Public Policy* 26(3), June: 151–61.

———. 2002. 'The Loss of Distance: Science in Transition'. In *Science, History, and Socialism: A Tribute to Everett Mendelsohn*, edited by Garland E. Allen and Roy M. MacLeod, 167–84. Dordrecht, Boston, and London: Kluwer Academic Publishers.

———. 2012. 'The Lure of the Mass Media and its Repercussions on Science'. In *The Sciences' Media Connection: Public Communication and its Repercussions*, edited by Simone Rödder, Martina Franzen, and Peter Weingart, 17–32. Dordrecht, Heidelberg, London, and New York: Springer.

Weingart, Peter and Petra Pansegrau. 1999. 'Reputation in Science and Prominence in the Media: The Goldhagen Debate'. *Public Understanding of Science* 8(1): 1–16.

Weingart, Peter, Anita Engels, and Petra Pansegrau. 2000. 'Risk Communication: Discourses on Climate Change in Science, Politics and Mass Media'. *Public Understanding of Science* 9(3): 261–83.

Wickramasinghe, Chandra. 2005. *A Journey with Fred Hoyle: The Search for Cosmic Life*, edited by Kamala Wickramasinghe. Singapore, Hackensack, and London: World Scientific Publishing Co. Pte. Ltd.

Wynne, Brian. 1989. 'Sheep Farming after Chernobyl: A Case Study in Communicating Scientific Information'. *Environment Magazine* 31(2): 10–15, 33–9.

———. 1995. 'Public Understanding of Science'. In *Handbook of Science and Technology Studies*, edited by Sheila Jasanoff et al., 361–88. London: Sage Publications.

———. 1996. 'Misunderstood Misunderstandings: Social Identities and the Public Uptake of Science'. In *Misunderstanding Science? The Public Reconstruction of Science and Technology*, edited by Alan Irwin and Brian Wynne, 19–46. Cambridge: Cambridge University Press.

———. 2003. 'Seasick on the Third Wave: Subverting the Hegemony of Propositionalism'. *Social Studies of Science* 33(3): 401–17.

———. 2007. 'Public Participation in Science and Technology: Performing and Obscuring a Political-Conceptual Category Mistake'. *East Asian Science, Technology, and Society: An International Journal* 1(1): 1–13.

Yamaguchi, Tomiko and Craig K. Harris. 2004. 'The Economic Homogenisation of Bt Cotton Discourse in India'. *Discourse and Society* 15(4): 467– 91.

Zachariah, Mathew and Sooryamoorthy. 1994. *Science for Social Revolution? Achievements and Dilemmas of a Development Movement: The Kerala Sastra Sahitya Parishad*. New Delhi: Visthaar Publications.

Zacharias, Usha. 2004. 'Intelligible Violence: Media Scripts, Hindu/Muslim Women, and the Battle for Citizenship in Kerala'. *Cultural Dynamics* 16(2–3): 169–92.

Ziman, John. 1996. '"Post Academic Science": Constructing Knowledge with Networks and Norms'. *Science Studies* 9(1): 67–80.

———. 2000. *Real Science: What It Is, and What It Means*. Cambridge: Cambridge University Press.

Filmography

Apothecary (2014, Malayalam), dir. Madhav Ramadasan, Arambankudiyil Cinemas, feature film, 138 min.

Dam 999 (2011, English), dir. Sohan Roy, BizTV Network, feature film, 108 min.

Dams: The Lethal Water Bombs (2011, English/dubbed into Malayalam), dir. Rahul Roy, Aries Telecasting Private Ltd., documentary film, 21 min.

Lingaa (2014, Tamil), dir. K.S. Ravikumar, Rockline Productions, feature film, 178 min.

Red Rain (2013, Malayalam), dir. Rahul Sadasivan, Highlands Entertainment, feature film, 101 min.

We Are the Aliens (14 November 2006, English), Horizon Science Show, BBC Two, television documentary, 50 min.

Index

British Army Engineering Corps, 127

capitalism, 38, 55, 62, 72n37
Cardiff Centre for Astrobiology, 178, 230
Cartoon(s), 207, 219n47, 219n48
caste(s), 53, 55, 65, 66, 247
casteist, 53
catastrophe discourse, 32, 128, 132
Central Groundwater Board, 142, 248n12
Central Marine Fisheries Research Institute (CMFRI), 180, 248n11
Central Plantation Crop Research Institute (CPCRI), 171
Central Soil and Material Research Station (CSMRS), 131
Centre for Earth Science Studies (CESS), 15, 18, 73, 119–23, 125, 129, 133, 135, 136, 139, 148n3, 155n85, 160n150, 164n184, 167–8, 169, 170, 175, 176, 184n4, 187n53, 187n63, 188n72, 198, 202, 207, 208, 209, 218n35, 221n64, 230, 237, 240, 248n11
    dispute over tremors and well collapses, 15, 128–30, 134–5, 202, 240
    scientist(s), 128, 130, 134, 136, 140, 142, 144, 154, 156n102, 158n115, 168, 169, 170, 182, 183, 186n51, 187n61, 188n72, 197, 199, 200, 210, 211, 212, 216n18, 216n27, 217n32, 218n39, 219n45, 222n69, 230, 249n17, 249n19
    scientists as villains, 170–7
Changanacherry, 69n6, 167, 169, 170, 175, 176, 180, 200, 218n39, 219n45
Cherama Doothan (magazine), 54

Chernobyl accident, 44n14
chief editor, 104n12, 109n83, 203
Chipko Andolan, 74n45
citizen groups, 28, 29
citizen science, 29, 45
citizen-collectives, 29, 45
citizen(s), 1, 6, 17, 24, 25, 29, 33, 36, 42, 45, 52, 56, 63, 65, 66, 67, 70n23, 139, 195, 241, 243, 244
citizen forums, 101, 233, 236
citizen jury (-ies), 29, 45n18
citizen-reader(s), 204, 205
citizenry, 62
citizenship, 71, 244–5, 247
citizenship claims, 247
citizenship, biological, 247
citizenship, scientific, 29, 247
civic epistemological frame(s), 160n137, 173, 176, 178, 183, 242, 243
civic epistemology, 30, 43, 46n22, 51, 65, 66, 67, 68, 68n1, 76n58, 138, 225, 242–3, 247, 249n23
civil society, 27, 35, 47n41, 48, 66, 72, 74, 78, 80, 90, 100, 101, 108n67, 209, 228, 236
Coca Cola, 245. *See also* Plachimada (struggle)
Cochin, 53, 55
Cochin Science Society, 71n30, 75n50
Cochin University of Science and Technology (CUSAT), 73n41, 173, 238
cold fusion controversy, 3, 220n55
coloured rain controversy, 166, 167, 176, 193, 197, 202, 210, 225–8, 229–30, 231, 233–4, 237–8
coloured rain(s), 15, 18, 141, 161, 166–90, 193, 194, 195, 196, 197, 201–2, 206, 210, 212, 216n20,

democratization of science. *See under* science

Department of Mining and Geology (DMG), 120, 238

Department of Science and Technology (DST), 21n23, 158n115, 239

Deshabhimani (newspaper), 11, 15, 22n26, 22n30, 55, 104n12, 107n50, 109n83, 121, 128, 131, 133, 134, 144, 145, 149n11, 149n14, 150n21, 156n102, 162n164, 163n184, 168, 169, 184n2, 216n23, 216n24, 219n48

Deshabhimani (magazine), 88

developmentalism, 124
  developmental crisis of Kuttanadu, 74
  developmental decision making, 6, 64
  developmental discourse, 6, 21n24, 61
  developmental disputes, 66
  developmental modernity, 6, 52, 64
  developmental planning, 65
  developmental projects, 51, 62–3, 245
  developmentalist state, 60, 62–3, 124, 144, 246
  Nehruvian. *See under* Nehru

Dewey, John, 49n53

Deweyian optimism, 40

diffusionist model, 28, 224. *See also* deficit model

diffusionist imagination of percolation of knowledge, 8

diffusionist-interventionist framework, 9

disaster experts, 193

discourse frame(s), 151n39, 200–2

discourse, risk, 4, 14, 27, 43, 64–7, 119. *See also* risks

domestication of science, 5, 57. *See also* science

Dravida Munnetra Kazhakam (DMK), 129, 153n71, 154n77

drug administration, 87, 91

Drug Controller General of India (DCGI), 78, 83, 87, 97, 107n50, 113n134, 239

drug research, 17, 78, 99–100, 103n3, 118, 225–6, 229–30, 237, 241
  Huang, Dr Ru Chih C., 78
  Roche, Ellen, 80
  drug testing on humans, 83, 90, 99, 103n3

Drugs and Cosmetic Rule of 1988, 90

dualism of nature and society, 26

earth science, 138, 140, 159n135

earth scientists, 121, 132, 134, 135, 140, 147, 171

earthquake, 118–19, 151n34, 152n52, 159n135, 197n4, 198n11, 198n15, 198n17, 200
  controversy, 122, 125, 128, 130, 167, 170, 193, 196, 214n4, 229–40
  dispute over the epicentre and intensity, 120–6, 130
  forecasts, 137–8, 156n108, 157, 198
  Gujarat, 119, 132, 135–42, 156n108, 158n115, 171, 215n13, 218n42, 219n44
  Idukki dam, 122–6, 133, 155n85
  Kerala, 118, 119, 124, 126, 207
  Koyna dam site, 124
  linkage with well collapses, 144
  low-intensity tremors, 140

Felt, Ulrich, 29
feminist politics, 246, 250n28
Fochler, Maximilian, 29
Food and Drug Administration
    (FDA), 80, 103n3
forced family planning, 63
foreshocks, 134, 140. *See also*
    earthquake
Forum for Patients' Rights (FPR), 86,
    108n67
fringe theories, 136, 138, 172, 183,
    206, 233
fumaroles, 172, 186n40
fuming-hill, 139, 162n159, 172,
    186n38, 195, 216n21, 218n41. *See
    also* Idinjimala
function systems theory, 31. *See also*
    Luhmann, Niklas; social systems;
    social (sub)system(s)
function systems, 24, 44n4
functional differentiation, 27, 44n4,
    46n28, 146
funding (of research), 2, 32, 80
funding agencies, 46n24, 239
fungal spores, 18, 174-6, 187n61,
    188n72

Gandhi, Indira, 63
Gandhi, M.K., 64
Gangadharan, Dr., 87, 88, 91, 96,
    111,114n148
gas emission, 157, 100, 172, 198n15
gate-keeping mechanism(s),
    160n137, 173, 176
genetically modified (GM) crops, 44n7
geological phenomena, 141,144-5,
    171, 176-7, 182, 194-6, 203,
    206, 212, 217n31, 229, 231, 239
geological structure, 143-4, 161, 169,
    235, 240
geomagnetic changes, 134

geopolitical dispute, 127-8
global science, 181
global research network, 100
globalization, 71n25
Goffman, Erving, 41
Gopalakrishnan, Dr K.R., 131, 136,
    149n11
governance, 26, 29, 36, 43, 88, 100,
    103, 118, 200, 227, 233, 238, 241,
    243
governance, risk, 26, 29, 36, 43, 88,
    103, 200, 238
governance, technological, 29, 43,
    242-3
government officials, 209
governmental interventions, 81-4, 234
Grandhasala Sangham, 55, 69n15
gravity dam, 127. *See also* dam
groundwater pressure, 142-5
groundwater pressure theory, 145,
    164n185. *See also* vanishing wells
Gujarat earthquake, the, 119, 132,
    135-42, 156n108, 158n115, 171,
    215n13, 218n42, 219n44. *See also*
    earthquake
Gupta, Dr Harsh, 215n12, 135

Habermas, Jurgen, 37-9, 48n44-7,
    49n48, 49n51-2, 50n56, 56, 57,
    70n24
Hagendijk, Rob, 35-6
hartals, 121, 149n13
Health Ministry Commission, 96, 98,
    113n135
Hindu,The (newspaper), 11-12,
    221n61, 223, 247n1
Hoyle, Fred, 177, 179, 189n82,
    189n85
Huang, Dr Ru Chih C., 78, 81, 84,
    89, 98, 105n24, 107n48. *See also*
    drug research

About the Author

Shiju Sam Varughese is Assistant Professor, Centre for Studies in Science, Technology and Innovation Policy, School of Social Sciences, Central University of Gujarat, Gandhinagar, India. His research is in the field of Science, Technology, and Society (STS) Studies. He is particularly interested in media and science communication, public engagement with science and technology, social history of knowledge, cultural studies of science and technology, and regional modernities in South Asia. He has edited (with Satheese Chandra Bose) *Kerala Modernity: Ideas, Spaces and Practices in Transition* (2015).